Foundations of Vector Retrieval

Sebastian Bruch

Foundations of Vector Retrieval

Springer

Sebastian Bruch
Pinecone
New York City, NY, USA

ISBN 978-3-031-55181-9 ISBN 978-3-031-55182-6 (eBook)
https://doi.org/10.1007/978-3-031-55182-6

This Springer imprint is published by the registered company Springer Nature Switzerland AG
The registered company address is: Gewerbestrasse 11, 6330 Cham, Switzerland

Paper in this product is recyclable.

Preface

We are witness to a few years of remarkable developments in Artificial Intelligence with the use of advanced machine learning algorithms, and in particular, *deep learning*. Gargantuan, complex neural networks that can learn through self-supervision—and quickly so with the aid of specialized hardware—transformed the research landscape so dramatically that, overnight it seems, many fields experienced not the usual, incremental progress, but rather a leap forward. Machine translation, natural language understanding, information retrieval, recommender systems, and computer vision are but a few examples of research areas that have had to grapple with the shock. Countless other disciplines beyond computer science such as robotics, biology, and chemistry too have benefited from deep learning.

These neural networks and their training algorithms may be complex, and the scope of their impact broad and wide, but nonetheless they are simply functions in a high-dimensional space. A trained neural network takes a *vector* as input, crunches and transforms it in various ways, and produces another vector, often in some other space. An image may thereby be turned into a vector, a song into a sequence of vectors, and a social network as a structured collection of vectors. It seems as though much of human knowledge, or at least what is expressed as text, audio, image, and video, has a vector representation in one form or another.

It should be noted that representing data as vectors is not unique to neural networks and deep learning. In fact, long before learnt vector representations of pieces of data—what is commonly known as "embeddings"—came along, data was often encoded as hand-crafted *feature* vectors. Each feature quantified into continuous or discrete values some facet of the data that was deemed relevant to a particular task (such as classification or regression). Vectors of that form, too, reflect our understanding of a real-world object or concept.

If new and old knowledge can be squeezed into a collection of learnt or hand-crafted vectors, what useful things does that enable us to do? A metaphor that might help us think about that question is this: An ever-evolving database full of such vectors that capture various pieces of data can be understood as a *memory* of sorts. We can then recall information from this memory to answer questions, learn about past and present events, reason about new problems, generate new content, and more.

Vector Retrieval

Mathematically, "recalling information" translates to finding vectors that are most *similar* to a *query* vector. The query vector represents what we wish to know more about, or recall information for. So, if we have a particular question in mind, the query is the vector representation of that question. If we wish to know more about an event, our query is that event expressed as a vector. If we wish to predict the function of a protein, perhaps we may learn a thing or two from known proteins that have a similar structure to the one in question, making a vector representation of the structure of our new protein a query.

Similarity is then a function of two vectors, quantifying how similar two vectors are. It may, for example, be based on the Euclidean distance between the query vector and a database vector, where similar vectors have a smaller distance. Or it may instead be based on the inner product between two vectors. Or their angle. Whatever function we use to measure similarity between pieces of data defines the structure of a database.

Finding k vectors from a database that have the highest similarity to a query vector is known as the top-k retrieval problem. When similarity is based on the Euclidean distance, the resulting problem is known as *nearest neighbor* search. Inner product for similarity leads to a problem known as *maximum inner product search*. Angular distance gives *maximum cosine similarity* search. These are mathematical formulations of the mechanism we called "recalling information."

The need to search for similar vectors from a large database arises in virtually every single one of our online transactions. Indeed, when we search the web for information about a topic, the search engine itself performs this similarity search over millions of web documents to find what may lexically or semantically match our query. Recommender systems find the most similar items to your browsing history by encoding items as vectors and, effectively, searching through a database of such items. Finding an old photo in a photo library, as another routine example, boils down to performing a similarity search over vector representations of images.

A neural network that is trained to perform a general task such as question-answering, could conceivably augment its view of the world by "recalling" information from such a database and finding answers to new questions. This is particularly useful for *generative* agents such as chatbots who would otherwise be frozen in time, and whose knowledge limited to what they were exposed to during their training. With a vector database on the side, however, they would have access to real-time information and can deduce new observations about content that is new to them. This is, in fact, the cornerstone of what is known as retrieval-augmented generation, an emerging learning paradigm.

Finding the most similar vectors to a query vector is easy when the database is small or when time is not of the essence: We can simply compare every vector in the database with the query and sort them by similarity. When the database grows large and the time budget is limited, as is often the case in practice, a naïve, exhaustive comparison of a query with database vectors is no longer realistic. That is where **vector retrieval** algorithms become relevant.

For decades now, research on vector retrieval has sought to improve the efficiency of search over large vector databases. The resulting literature is rich with solutions ranging from heavily theoretical results to performant empirical heuristics. Many of the proposed algorithms have undergone rigorous benchmarking and have been challenged in competitions at major conferences. Technology giants and startups alike have invested heavily in developing open-source libraries and managed infrastructure that offer fast and scalable vector retrieval.

That is not the end of that story, however. Research continues to date. In fact, how we do vector retrieval today faces a stress-test as databases grow orders of magnitude larger than ever before. None of the existing methods, for example, proves easy to scale to a database of billions of high-dimensional vectors, or a database whose records change frequently.

About This Monograph

The need to conduct more research underlines the importance of making the existing literature more readily available and the research area more inviting. That is partially fulfilled with existing surveys that report the state of the art at various points in time. However, these publications are typically focused on a single class of vector retrieval algorithms, and compare and contrast published methods by their empirical performance alone. Importantly, no manuscript has yet summarized major algorithmic milestones in the vast vector retrieval literature, or has been prepared to serve as a reference for new and established researchers.

That gap is what this monograph intends to close. With the goal of presenting the fundamentals of vector retrieval as a sub-discipline, this manuscript delves into important data structures and algorithms that have emerged in the literature to solve the vector retrieval problem efficiently and effectively.

Structure

This monograph is divided into four parts. The first part introduces the problem of vector retrieval and formalizes the concepts involved. The second part delves into retrieval algorithms that help solve the vector retrieval problem efficiently and effectively. Part three is devoted to vector compression. Finally, the fourth part presents a review of background material in a series of appendices.

Introduction

We start with a thorough **introduction** to the problem itself in Chapter 1 where we define the various flavors of vector retrieval. We then elaborate what is so difficult about the problem in high-dimensional spaces in Chapter 2.

In fact, sometimes high-dimensional spaces are hopeless. However, in reality data often lie on some low-dimensional space, even though their naïve vector representations are in high dimensions. In those cases, it turns out, we can do much better. Exactly how we characterize this low **"intrinsic" dimensionality** is the topic of Chapter 3.

Retrieval Algorithms

With that foundation in place and the question clearly formulated, the second part of the monograph explores the different classes of existing solutions in great depth. We close each chapter with a summary of algorithmic insights. There, we will also discuss what remains challenging and explore future research directions.

We start with **branch-and-bound** algorithms in Chapter 4. The high-level idea is to lay a hierarchical mesh over the space, then given a query point navigate the hierarchy to the cell that likely contains the solution. We will see, however, that in high dimensions, the basic forms of these methods become highly inefficient to the point where an exhaustive search likely performs much better.

Alternatively, instead of laying a mesh over the space, we may define a fixed number of buckets and map data points to these buckets with the property that, if two data points are close to each other according to the distance function, they are more likely to be mapped to the same bucket. When processing a query, we find which bucket it maps to and search the data points in that bucket. This is the intuition that led to the family of **Locality Sensitive Hashing** (LSH) algorithms—a topic we will discuss in depth in Chapter 5.

Yet another class of ideas adopts the view that data points are nodes in a graph. We place an edge between two nodes if they are among each others' nearest neighbors. When presented with a query point, we enter the graph through one of the nodes and greedily traverse the edges by taking the edge that leads to the minimum distance with the query. This process is repeated until we are stuck in some (local) optima. This is the core idea in **graph** algorithms, as we will learn in Chapter 6.

The final major approach is the simplest of all: Organize the data points into small clusters during pre-processing. When a query point arrives, solve the "cluster retrieval" problem first, then solve retrieval on the chosen clusters. We will study this **clustering** method in detail in Chapter 7.

As we examine vector retrieval algorithms, it is inevitable that we must ink in extra pages to discuss why similarity based on inner product is special and why it poses extra challenges for the algorithms in each category—many of these difficulties will become clear in the introductory chapters.

There is, however, a special class of algorithms specifically for inner product. **Sampling** algorithms take advantage of the linearity of inner product to reduce the dependence of the time complexity on the number of dimensions. We will review example algorithms in Chapter 8.

Compression

The third part of this monograph concerns the storage of vectors and their distance computation. After all, the vector retrieval problem is not just concerned with the time complexity of the retrieval process itself, but also aims to reduce the size of the data structure that helps answer queries—known as the index. Compression helps that cause.

In Chapter 9 we will review how vectors can be **quantized** to reduce the size of the index while simultaneously facilitating fast computation of the distance function in the compressed domain! That is what makes quantization effective but challenging.

Related to the topic of compression is the concept of **sketching**. Sketching is a technique to project a high-dimensional vector into a low-dimensional vector, called a *sketch*, such that certain properties (e.g., the L_2 norm, or inner products between any two vectors) are *approximately* preserved. This probabilistic method of reducing dimensionality naturally connects to vector retrieval. We offer a peek into the vast sketching literature in Chapter 10 and discuss its place in the vector retrieval research. We do so with a particular focus on *sparse* vectors in an inner product space—contrasting sketching with quantization methods that are more appropriate for *dense* vectors.

Objective

It is important to stress, however, that the purpose of this monograph is *not* to provide a comprehensive survey or comparative analysis of every published work that has appeared in the vector retrieval literature. There is simply too many empirical works with volumes of heuristics and engineering solutions to cover. Instead, we will give an in-depth, didactic treatment of foundational ideas that have caused a seismic shift in how we approach the problem, and the theory that underpins them.

By consolidating these ideas, this monograph hopes to make this fascinating field more inviting—especially to the uninitiated—and enticing as a research topic to new and established researchers. We hope the reader will find that this monograph delivers on these objectives.

Intended Audience

This monograph is intended as an introductory text for graduate students who wish to embark on research on vector retrieval. It is also meant to serve as a self-contained reference that captures important developments in the field, and as such, may be useful to established researchers as well.

As the work is geared towards researchers, however, it naturally emphasizes the theoretical aspects of algorithms as opposed to their empirical behavior or experimental performance. We present theorems and their proofs, for example. We do not, on the other hand, present experimental results or compare algorithms on datasets systematically. There is also no discussion around the use of the presented algorithms in practice, notes on implementation and libraries, or practical insights and heuristics that are often critical to making these algorithms work on real data. As a result, practitioners or applied researchers may not find the material immediately relevant.

Finally, while we make every attempt to articulate the theoretical results and explain the proofs thoroughly, having some familiarity with linear algebra and probability theory helps digest the results more easily. We have included a review of the relevant concepts and results from these subjects in Appendices B (probability), C (concentration inequalities), and D (linear algebra) for convenience. Should the reader wish to skip the proofs, however, the narrative should still paint a complete picture of how each algorithm works.

Acknowledgements

I am forever indebted to my dearest colleagues Edo Liberty, Amir Ingber, Brian Hentschel, and Aditya Krishnan. This incredible but humble group of scholars at Pinecone are generous with their time and knowledge, patiently teaching me what I do not know, and letting me use them as a sounding board without fail. Their encouragement throughout the process of writing this manuscript, too, was the force that drove this work to completion.

I am also grateful to Claudio Lucchese, a dear friend, a co-author, and a professor of computer science at the Ca' Foscari University of Venice, Italy. I conceived of the idea for this monograph as I lectured at Ca' Foscari on the topic of retrieval and ranking, upon Claudio's kind invitation.

I would not be writing these words were it not for the love, encouragement, and wisdom of Franco Maria Nardini, of the ISTI CNR in Pisa, Italy. In the mad and often maddening world of research, Franco is the one knowledgeable and kind soul who restores my faith in research and guides me as I navigate the landscape.

Finally, there are no words that could possibly convey my deepest gratitude to my partner, Katherine, for always supporting me and my ambitions; for showing by example what dedication, tenacity, and grit ought to mean; and for finding me when I am lost.

Notation

This section summarizes the special symbols and notation used throughout this work. We often repeat these definitions in context as a reminder, especially if we choose to abuse notation for brevity or other reasons.

Paragraphs that are highlighted in a gray box such as this contain important statements, often conveying key findings or observations, or a detail that will be important to recall in later chapters.

Terminology

We use the terms "vector" and "point" interchangeably. In other words, we refer to an ordered list of d real values as a d-dimensional vector or a point in $\mathbb{R}^d$.

We say that a point is a *data* point if it is part of the collection of points we wish to sift through. It is a *query* point if it is the input to the search procedure, and for which we are expected to return the top-k similar data points from the collection.

Symbols

Reserved Symbols

$\mathcal{X}$	Used exclusively to denote a collection of vectors.
m	We use this symbol exclusively to denote the cardinality of a collection of data points, $\mathcal{X}$.
q	Used singularly to denote a query point.
d	We use this symbol exclusively to refer to the number of dimensions.
$e_1, e_2, \ldots, e_d$	Standard basis vectors in $\mathbb{R}^d$

Sets

$\mathcal{J}$	Calligraphic font typically denotes sets.
$\lvert\cdot\rvert$	The cardinality (number of items) of a finite set.
$[n]$	The set of integers from 1 to n (inclusive): $\{1, 2, 3, \ldots, n\}$.
$B(u, r)$	The closed ball of radius r centered at point u: $\{v \mid \delta(u, v) \leq r\}$ where $\delta(\cdot, \cdot)$ is the distance function.
$\setminus$	The set difference operator: $\mathcal{A} \setminus \mathcal{B} = \{x \in \mathcal{A} \mid x \notin \mathcal{B}\}$.
$\triangle$	The symmetric difference of two sets.
$\mathbb{1}_p$	The indicator function. It is 1 if the predicate p is true, and 0 otherwise.

Vectors and Vector Space

$[a, b]$	The closed interval from a to b.
$\mathbb{Z}$	The set of integers.
$\mathbb{R}^d$	d-dimensional Euclidean space.
$\mathbb{S}^{d-1}$	The hypersphere in $\mathbb{R}^d$.
u, v, w	Lowercase letters denote vectors.
u_i, v_i, w_i	Subscripts identify a specific coordinate of a vector, so that u_i is the i-th coordinate of vector u.

Functions and Operators

$nz(\cdot)$	The set of non-zero coordinates of a vector: $nz(u) = \{i \mid u_i \neq 0\}$.
$\delta(\cdot, \cdot)$	We use the symbol δ exclusively to denote the distance function, taking two vectors and producing a real value.
$J(\cdot, \cdot)$	The Jaccard similarity index of two vectors: $J(u, v) = \lvert nz(u) \cap nz(v)\rvert / \lvert nz(u) \cup nz(v)\rvert$.
$\langle \cdot, \cdot \rangle$	Inner product of two vectors: $\langle u, v \rangle = \sum_i u_i v_i$.
$\lVert\cdot\rVert_p$	The L_p norm of a vector: $\lVert u \rVert_p = (\sum_i \lvert u_i \rvert^p)^{1/p}$.
$\oplus$	The concatenation of two vectors. If $u, v \in \mathbb{R}^d$, then $u \oplus v \in \mathbb{R}^{2d}$.

Probabilities and Distributions

$\mathbb{E}[\cdot]$	The expected value of a random variable.
$\mathrm{Var}[\cdot]$	The variance of a random variable.
$\mathbb{P}[\cdot]$	The probability of an event.
$\wedge$, $\vee$	Logical AND and OR operators.
Z	We generally use uppercase letters to denote random variables.

Contents

Part II Retrieval Algorithms

Part I
Introduction

Chapter 1
Vector Retrieval

Abstract This chapter sets the stage for the remainder of this monograph. It explains where vectors come from, how they have come to represent data of any modality, and why they are a useful mathematical tool in machine learning. It then describes the structure we typically expect from a collection of vectors: that similar objects get vector representations that are close to each other in an inner product or metric space. We then define the problem of top-k retrieval over a well-structured collection of vectors, and explore its different flavors, including approximate retrieval.

1.1 Vector Representations

We routinely use ordered lists of numbers, or *vectors*, to describe objects of any shape or form. Examples abound. Any geographic location on earth can be recognized as a vector consisting of its latitude and longitude. A desk can be described as a vector that represents its dimensions, area, color, and other quantifiable properties. A photograph as a list of pixel values that together paint a picture. A sound wave as a sequence of frequencies.

Vector representations of objects have long been an integral part of the machine learning literature. Indeed, a classifier, a regression model, or a ranking function learns patterns from, and acts on, vector representations of data. In the past, this vector representation of an object was nothing more than a collection of its *features*. Every feature described some facet of the object (for example, the color intensity of a pixel in a photograph) as a continuous or discrete value. The idea was that, while individual features describe only a small part of the object, together they provide sufficiently powerful statistics about the object and its properties for the machine learnt model to act on.

The features that led to the vector representation of an object were generally hand-crafted functions. To make sense of that, let us consider a text document in English. Strip the document of grammar and word order, and

S. Bruch, *Foundations of Vector Retrieval*, https://doi.org/10.1007/978-3-031-55182-6_1

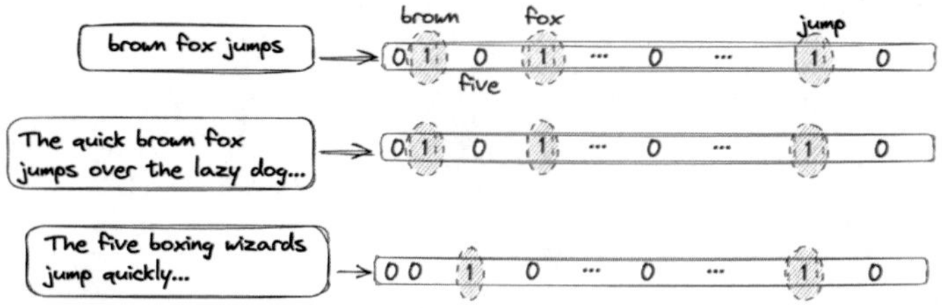

Fig. 1.1: Vector representation of a piece of text by adopting a "bag of words" view: A text document, when stripped of grammar and word order, can be thought of as a vector, where each coordinate represents a term in our vocabulary and its value records the frequency of that term in the document or some function of it. The resulting vectors are typically *sparse*; that is, they have very few non-zero coordinates.

we end up with a *set* of words, more commonly known as a "bag of words." This set can be summarized as a histogram.

If we designated every term in the English vocabulary to be a dimension in a (naturally) high-dimensional space, then the histogram representation of the document can be encoded as a vector. The resulting vector has relatively few non-zero coordinates, and each non-zero coordinate records the frequency of a term present in the document. This is illustrated in Figure 1.1 for a toy example. More generally, non-zero values may be a function of a term's frequency in the document and its propensity in a collection—that is, the likelihood of encountering the term [Salton and Buckley, 1988].

The advent of *deep learning* and, in particular, Transformer-based models [Vaswani et al., 2017] brought about vector representations that are beyond the elementary formation above. The resulting representation is often, as a single entity, referred to as an *embedding*, instead of a "feature vector," though the underlying concept remains unchanged: an object is encoded as a real d-dimensional vector, a point in $\mathbb{R}^d$.

Let us go back to the example from earlier to see how the embedding of a text document could be different from its representation as a frequency-based feature vector. Let us maintain the one-to-one mapping between coordinates and terms in the English vocabulary. Remember that in the "lexical" representation from earlier, if a coordinate was non-zero, that implied that the corresponding term was present in the document and its value indicated its frequency-based feature. Here we instead *learn* to turn coordinates on or off and, when we turn a coordinate on, we want its value to predict the significance of the corresponding term based on semantics and contextual information. For example, the (absent) synonyms of a (present) term may get a non-zero value, and terms that offer little discriminative power in the given context become 0 or close to it. This basic idea has been explored extensively by many recent models of text [Bai et al., 2020, Formal et al., 2021, 2022, Zhuang and Zuccon, 2022, Dai and Callan, 2020, Gao et al., 2021, Mallia et al., 2021, Zamani et al., 2018, Lin and Ma, 2021] and has been shown to produce effective representations.

Vector representations of text need not be sparse. While sparse vectors with dimensions that are grounded in the vocabulary are inherently *interpretable*, text documents can also be represented with lower-dimensional *dense* vectors (where every coordinate is *almost surely* non-zero). This is, in fact, the most dominant form of vector representation of text documents in the literature [Lin et al., 2021, Karpukhin et al., 2020, Xiong et al., 2021, Reimers and Gurevych, 2019, Santhanam et al., 2022, Khattab and Zaharia, 2020]. Researchers have also explored *hybrid* representations of text where vectors have a dense subspace and an orthogonal sparse subspace [Chen et al., 2022, Bruch et al., 2023, Wang et al., 2021, Kuzi et al., 2020, Karpukhin et al., 2020, Ma et al., 2021, 2020, Wu et al., 2019].

Unsurprisingly, the same embedding paradigm can be extended to other data modalities beyond text: Using deep learning models, one may embed images, videos, and audio recordings into vectors. In fact, it is even possible to project different data modalities (e.g., images and text) together into the same vector space and preserve some property of interest [Zhang et al., 2020, Guo et al., 2019].

It appears, then, that vectors are everywhere. Whether they are the result of hand-crafted functions that capture features of the data or are the output of learnt models; whether they are dense, sparse, or both, they are effective representations of data of any modality.

But what precisely is the point of turning every piece of data into a vector? One answer to that question takes us to the fascinating world of *retrieval*.

1.2 Vectors as Units of Retrieval

It would make for a vapid exercise if all we had were vector representations of data without any structure governing a collection of them. To give a collection of points some structure, we must first ask ourselves what goal we are trying to achieve by turning objects into vectors. It turns out, we often intend for the vector representation of two *similar* objects to be "close" to each other according to some well-defined distance function.

That is the structure we desire: Similarity in the vector space must imply similarity between objects. So, as we engineer features to be extracted from an object, or design a protocol to learn a model to produce embeddings of data, we must choose the dimensionality d of the target space (a subset of $\mathbb{R}^d$)

along with a distance function $\delta(\cdot, \cdot)$. Together, these define an inner product or metric space.

Consider again the lexical representation of a text document where d is the size of the English vocabulary. Let δ be the distance variant of the Jaccard index, $\delta(u, v) = -J(u, v) \triangleq -|nz(u) \cap nz(v)|/|nz(u) \cup nz(v)|$, where $nz(u) = \{i \mid u_i \neq 0\}$ with u_i denoting the i-th coordinate of vector u.

In the resulting space, if vectors u and v have a smaller distance than vectors u and w, then we can clearly conclude that the document represented by u is lexically more similar to the one represented by v than it is to the document w represents. That is because the distance (or, in this case, similarity) function reflects the amount of overlap between the terms present in one document with another.

We should be able to make similar arguments given a semantic embedding of text documents. Again consider the sparse embeddings with d being the size of the vocabulary, and more concretely, take Splade [Formal et al., 2021] as a concrete example. This model produces real-valued sparse vectors in an inner product space. In other words, the objective of its learning procedure is to maximize the inner product between similar vectors, where the inner product between two vectors u and v is denoted by $\langle u, v \rangle$ and is computed using $\sum_i u_i v_i$.

In the resulting space, if u, v, and w are generated by Splade with the property that $\langle u, v \rangle > \langle u, w \rangle$, then we can conclude that, according to Splade, documents represented by u and v are semantically more similar to each other than u is to w. There are numerous other examples of models that optimize for the angular distance or Euclidean distance (L_2) between vectors to preserve (semantic) similarity.

What can we do with a well-characterized collection of vectors that represent real-world objects? Quite a lot, it turns out. One use case is the topic of this monograph: the fundamental problem of retrieval.

> We are often interested in finding k objects that have the highest degree of similarity to a query object. When those objects are represented by vectors in a collection $\mathcal{X}$, where the distance function $\delta(\cdot, \cdot)$ is reflective of similarity, we may formalize this top-k question mathematically as finding the k minimizers of distance with the query point!

We state that formally in the following definition:

Definition 1.1 (Top-k Retrieval) Given a distance function $\delta(\cdot, \cdot)$, we wish to pre-process a collection of data points $\mathcal{X} \subset \mathbb{R}^d$ in time that is polynomial in $|\mathcal{X}|$ and d, to form a data structure (the "index") whose size is polynomial in $|\mathcal{X}|$ and d, so as to efficiently solve the following in time $o(|\mathcal{X}|d)$ for an arbitrary query $q \in \mathbb{R}^d$:

$$\underset{u \in \mathcal{X}}{\arg\min}^{(k)}\, \delta(q, u). \tag{1.1}$$

A web search engine, for example, finds the most relevant documents to your query by first formulating it as a top-k retrieval problem over a collection of (not necessarily text-based) vectors. In this way, it quickly finds the subset of documents from the entire web that may satisfy the information need captured in your query. Question answering systems, conversational agents (such as Siri, Alexa, and ChatGPT), recommendation engines, image search, outlier detectors, and myriad other applications that are at the forefront of many online services and in many consumer gadgets depend on data structures and algorithms that can answer the top-k retrieval question as efficiently and as effectively as possible.

1.3 Flavors of Vector Retrieval

We create an instance of the deceptively simple problem formalized in Definition 1.1 the moment we acquire a collection of vectors $\mathcal{X}$ together with a distance function δ. In the remainder of this monograph, we assume that there is some function, either manually engineered or learnt, that transforms objects into vectors. So, from now on, $\mathcal{X}$ is a given.

The distance function then, specifies the flavor of the top-k retrieval problem we need to solve. We will review these variations and explore what each entails.

1.3.1 Nearest Neighbor Search

In many cases, the distance function is derived from a proper metric where non-negativity, coincidence, symmetry, and triangle inequality hold for δ. A clear example of this is the L_2 distance: $\delta(u, v) = \|u - v\|_2$. The resulting problem, illustrated for a toy example in Figure 1.2(a), is known as k-Nearest Neighbors (k-NN) search:

$$\underset{u \in \mathcal{X}}{\arg\min}^{(k)} \|q - u\|_2 = \underset{u \in \mathcal{X}}{\arg\min}^{(k)} \|q - u\|_2^2. \tag{1.2}$$

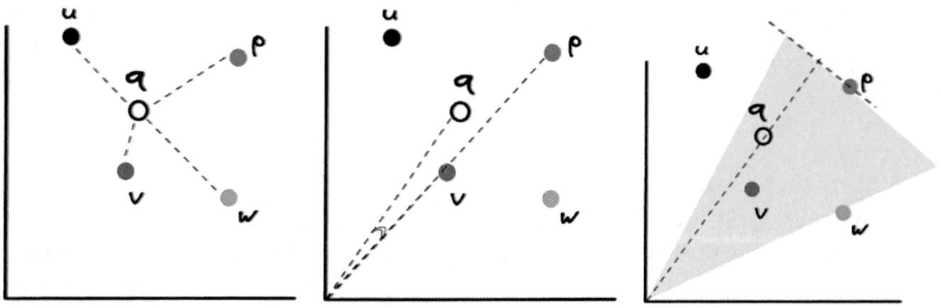

(a) k-NN (b) k-MCS (c) k-MIPS

Fig. 1.2: Variants of vector retrieval for a toy vector collection in $\mathbb{R}^2$. In Nearest Neighbor search, we find the data point whose L_2 distance to the query point is minimal (v for top-1 search). In Maximum Cosine Similarity search, we instead find the point whose angular distance to the query point is minimal (v and p are equidistant from the query). In Maximum Inner Product Search, we find a vector that maximizes the inner product with the query vector. This can be understood as letting the hyperplane orthogonal to the query point sweep the space towards the origin; the first vector to touch the sweeping plane is the maximizer of inner product. Another interpretation is this: the shaded region in the figure contains all the points y for which p is the answer to $\arg\max_{x \in \{u,v,w,p\}} \langle x, y \rangle$.

1.3.2 Maximum Cosine Similarity Search

The distance function may also be the angular distance between vectors, which is again a proper metric. The resulting minimization problem can be stated as follows, though its equivalent maximization problem (involving the cosine of the angle between vectors) is perhaps more recognizable:

$$\underset{u \in \mathcal{X}}{\overset{(k)}{\arg\min}}\ 1 - \frac{\langle q, u \rangle}{\|q\|_2 \|u\|_2} = \underset{u \in \mathcal{X}}{\overset{(k)}{\arg\max}}\ \frac{\langle q, u \rangle}{\|u\|_2}. \tag{1.3}$$

The latter is referred to as the k-Maximum Cosine Similarity (k-MCS) problem. Note that, because the norm of the query point, $\|q\|_2$, is a constant in the optimization problem, it can simply be discarded; the resulting distance function is rank-equivalent to the angular distance. Figure 1.2(b) visualizes this problem on a toy collection of vectors.

1.3.3 Maximum Inner Product Search

Both of the problems in Equations (1.2) and (1.3) are special instances of a more general problem known as k-Maximum Inner Product Search (k-MIPS):

$$\underset{u\in\mathcal{X}}{\arg\max^{(k)}}\langle q,u\rangle. \tag{1.4}$$

This is easy to see for k-MCS: If, in a pre-processing step, we L_2-normalized all vectors in $\mathcal{X}$ so that u is transformed to $u' = u/\|u\|_2$, then $\|u'\|_2 = 1$ and therefore Equation (1.3) reduces to Equation (1.4).

As for a reduction of k-NN to k-MIPS, we can expand Equation (1.2) as follows:

$$\begin{aligned}\underset{u\in\mathcal{X}}{\arg\min^{(k)}}\|q-u\|_2^2 &= \underset{u\in\mathcal{X}}{\arg\min^{(k)}}\|q\|_2^2 - 2\langle q,u\rangle + \|u\|_2^2\\ &= \underset{u'\in\mathcal{X}'}{\arg\max^{(k)}}\langle q',u'\rangle,\end{aligned}$$

where we have discarded the constant term, $\|q\|_2^2$, and defined $q' \in \mathbb{R}^{d+1}$ as the concatenation of $q \in \mathbb{R}^d$ and a 1-dimensional vector with value $-1/2$ (i.e., $q' = [q, -1/2]$), and $u' \in \mathbb{R}^{d+1}$ as $[u, \|u\|_2^2]$.

The k-MIPS problem, illustrated on a toy collection in Figure 1.2(c), does not come about just as the result of the reductions shown above. In fact, there exist embedding models (such as SPLADE, as discussed earlier) that learn vector representations with respect to inner product as the distance function. In other words, k-MIPS is an important problem in its own right.

1.3.3.1 Properties of MIPS

In a sense, then, it is sufficient to solve the k-MIPS problem as it is the umbrella problem for much of vector retrieval. Unfortunately, k-MIPS is a much harder problem than the other variants. That is because inner product is not a proper metric. In particular, it is not non-negative and does not satisfy the triangle inequality, so that $\langle u,v\rangle \not\leq \langle u,w\rangle + \langle w,v\rangle$ in general.

Perhaps more troubling is the fact that even "coincidence" is not guaranteed. In other words, it is not true in general that a vector u maximizes inner product with itself: $u \neq \arg\max_{v\in\mathcal{X}}\langle v,u\rangle$!

As an example, suppose v and $p = \alpha v$ for some $\alpha > 1$ are vectors in the collection $\mathcal{X}$—a case demonstrated in Figure 1.2(c). Clearly, we have that

$\langle v, p \rangle = \alpha \langle v, v \rangle > \langle v, v \rangle$, so that p (and not v) is the solution to MIPS[1] for the query point v.

In high-enough dimensions and under certain statistical conditions, however, coincidence is reinstated for MIPS with high probability. One such case is stated in the following theorem.

Theorem 1.1 *Suppose data points $\mathcal{X}$ are independent and identically distributed (*iid*) in each dimension and drawn from a zero-mean distribution. Then, for any $u \in \mathcal{X}$:*

$$\lim_{d \to \infty} \mathbb{P}\left[u = \underset{v \in \mathcal{X}}{\arg\max} \langle u, v \rangle\right] = 1.$$

Proof. Denote by $\mathrm{Var}[\cdot]$ and $\mathbb{E}[\cdot]$ the variance and expected value operators. By the conditions of the theorem, it is clear that $\mathbb{E}[\langle u, u \rangle] = d\,\mathbb{E}[Z^2]$ where Z is the random variable that generates each coordinate of the vector. We can also see that $\mathbb{E}[\langle u, X \rangle] = 0$ for a random data point X, and that $\mathrm{Var}[\langle u, X \rangle] = \|u\|_2^2\,\mathbb{E}[Z^2]$.

We wish to claim that $u \in \mathcal{X}$ is the solution to a MIPS problem where u is also the query point. That happens if and only if every other vector in $\mathcal{X}$ has an inner product with u that is smaller than $\langle u, u \rangle$. So that:

$$\begin{aligned}
\mathbb{P}\left[u = \underset{v \in \mathcal{X}}{\arg\max} \langle u, v \rangle\right] &= \mathbb{P}\left[\langle u, v \rangle \leq \langle u, u \rangle \quad \forall\, v \in \mathcal{X}\right] = \\
&1 - \mathbb{P}\left[\exists\, v \in \mathcal{X} \ s.t. \quad \langle u, v \rangle > \langle u, u \rangle\right] \geq && \text{(by Union Bound)} \\
&1 - \sum_{v \in \mathcal{X}} \mathbb{P}\left[\langle u, v \rangle > \langle u, u \rangle\right] = && \text{(by } iid\text{)} \\
&1 - |\mathcal{X}|\,\mathbb{P}\left[\langle u, X \rangle > \langle u, u \rangle\right].
\end{aligned}$$

Let us turn to the last term and bound the probability for a random data point:

$$\mathbb{P}\left[\langle u, X \rangle > \langle u, u \rangle\right] = \mathbb{P}\left[\underbrace{\langle u, X \rangle - \langle u, u \rangle + d\,\mathbb{E}[Z^2]}_{Y} > d\,\mathbb{E}[Z^2]\right].$$

The expected value of Y is 0. Denote by σ^2 its variance. By the application of the one-sided Chebyshev's inequality,[2] we arrive at the following bound:

$$\mathbb{P}\left[\langle u, X \rangle > \langle u, u \rangle\right] \leq \frac{\sigma^2}{\sigma^2 + d^2\,\mathbb{E}[Z^2]^2}.$$

---

[1] When $k = 1$, we drop the symbol k from the name of the retrieval problem. So we write MIPS instead of 1-MIPS.

[2] The one-sided Chebyshev's inequality for a random variable X with mean μ and variance σ^2 states that $\mathbb{P}\left[X - \mu > t\right] \leq \sigma^2/(\sigma^2 + t^2)$.

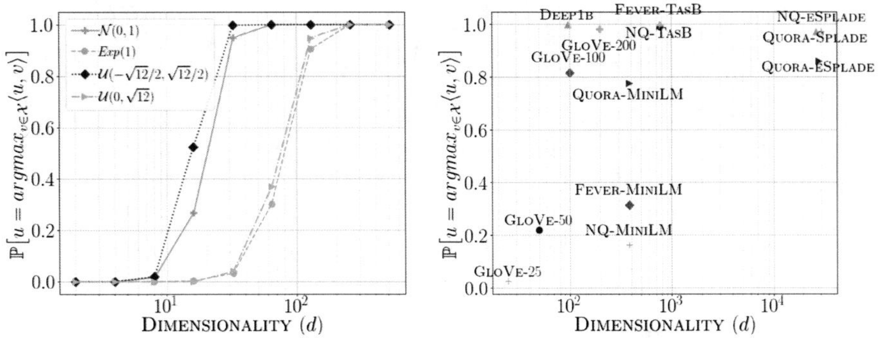

(a) Synthetic (b) Real

Fig. 1.3: Probability that $u \in \mathcal{X}$ is the solution to MIPS over $\mathcal{X}$ with query u versus the dimensionality d, for various synthetic and real collections $\mathcal{X}$. For synthetic collections, $|\mathcal{X}| = 100{,}000$. Appendix A gives a description of the real collections. Note that, for real collections, we estimate the reported probability by sampling 10,000 data points and using them as queries. Furthermore, we do not pre-process the vectors—importantly, we do not L_2-normalize the collections.

Note that, σ^2 is a function of the sum of *iid* random variables, and, as such, grows linearly with d. In the limit. this probability tends to 0. We have thus shown that $\lim_{d \to \infty} \mathbb{P}\left[u = \arg\max_{v \in \mathcal{X}} \langle u, v \rangle\right] \geq 1$ which concludes the proof. □

1.3.3.2 Empirical Demonstration of the Lack of Coincidence

Let us demonstrate the effect of Theorem 1.1 empirically. First, let us choose distributions that meet the requirements of the theorem: a Gaussian distribution with mean 0 and variance 1, and a uniform distribution over $[-\sqrt{12}/2, \sqrt{12}/2]$ (with variance 1) will do. For comparison, choose another set of distributions that do not have the requisite properties: Exponential with rate 1 and uniform over $[0, \sqrt{12}]$. Having fixed the distributions, we next sample 100,000 random vectors from them to form a collection $\mathcal{X}$. We then take each data point, use it as a query in MIPS over $\mathcal{X}$, and report the proportion of data points that are solutions to their own search.

Figure 1.3(a) illustrates the results of this experiment. As expected, for the Gaussian and centered uniform distributions, the ratio of interest approaches 1 when d is sufficiently large. Surprisingly, even when the distributions do not strictly satisfy the conditions of the theorem, we still observe the convergence

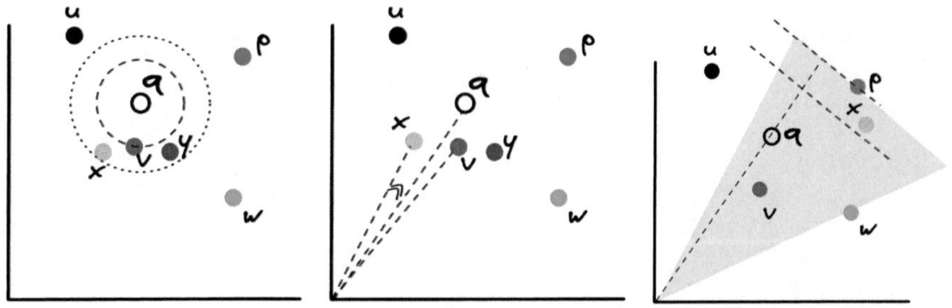

Fig. 1.4: Approximate variants of top-1 retrieval for a toy collection in $\mathbb{R}^2$. In NN, we admit vectors that are at most ϵ away from the optimal solution. As such, x and y are both valid solutions as they are in a ball with radius $(1+\epsilon)\delta(q,x)$ centered at the query. Similarly, in MCS, we accept a vector (e.g., x) if its angle with the query point is at most $1+\epsilon$ greater than the angle between the query and the optimal vector (i.e., v). For the MIPS example, assuming that the inner product of query and x is at most $(1-\epsilon)$-times the inner product of query and p, then x is an acceptable solution.

of that ratio to 1. So it appears that the requirements of Theorem 1.1 are more forgiving than one may imagine.

We also repeat the exercise above on several real-world collections, a description of which can be found in Appendix A along with salient statistics. The results of these experiments are visualized in Figure 1.3(b). As expected, whether a data point maximizes inner product with itself entirely depends on the underlying data distribution. We can observe that, for some collections in high dimensions, we are likely to encounter coincidence in the sense we defined earlier, but for others that is clearly not the case. It is important to keep this difference between synthetic and real collections in mind when designing experiments that evaluate the performance of MIPS systems.

1.4 Approximate Vector Retrieval

Saying one problem is harder than another neither implies that we cannot approach the harder problem, nor does it mean that the "easier" problem is easy to solve. In fact, none of these variants of vector retrieval (k-NN, k-MCS, and k-MIPS) can be solved exactly *and* efficiently in high dimensions. Instead, we must either accept that the solution would be inefficient (in terms of space- or time-complexity), or allow some degree of error.

The first case of solving the problem exactly but inefficiently is uninteresting: If we are looking to find the solution for $k = 1$, for example, it is enough to compute the distance function for every vector in the collection and the query, resulting in linear complexity. When $k > 1$, the total time complexity is $\mathcal{O}(|\mathcal{X}|d \log k)$, where $|\mathcal{X}|$ is the size of the collection. So it typically makes more sense to investigate the second strategy of admitting error.

That argument leads naturally to the class of ϵ-*approximate* vector retrieval problems. This idea can be formalized rather easily for the special case where $k = 1$: The approximate solution for the top-1 retrieval is satisfactory so long as the vector u returned by the algorithm is at most $(1 + \epsilon)$ factor farther than the optimal vector u^*, according to $\delta(\cdot, \cdot)$ and for some arbitrary $\epsilon > 0$:

$$\delta(q, u) \leq (1 + \epsilon)\delta(q, u^*). \tag{1.5}$$

Figure 1.4 renders the solution space for an example collection in $\mathbb{R}^2$.

The formalism above extends to the more general case where $k > 1$ in an obvious way: a vector u is a valid solution to the ϵ-approximate top-k problem if its distance to the query point is at most $(1 + \epsilon)$ times the distance to the k-th optimal vector. This is summarized in the following definition:

Definition 1.2 (ϵ-Approximate Top-k Retrieval) Given a distance function $\delta(\cdot, \cdot)$, we wish to pre-process a collection of data points $\mathcal{X} \subset \mathbb{R}^d$ in time that is polynomial in $|\mathcal{X}|$ and d, to form a data structure (the "index") whose size is polynomial in $|\mathcal{X}|$ and d, so as to efficiently solve the following in time $o(|\mathcal{X}|d)$ for an arbitrary query $q \in \mathbb{R}^d$ and $\epsilon > 0$:

$$\mathcal{S} = \underset{u \in \mathcal{X}}{\overset{(k)}{\arg\min}}\, \delta(q, u),$$

such that for all $u \in \mathcal{S}$, Equation (1.5) is satisfied where u^* is the k-th optimal vector obtained by solving the problem in Definition 1.1.

Despite the extension to top-k above, it is more common to characterize the effectiveness of an approximate top-k solution as the percentage of correct vectors that are present in the solution. Concretely, if $\mathcal{S} = \arg\max^{(k)}_{u \in \mathcal{X}} \delta(q, u)$ is the exact set of top-k vectors, and $\tilde{\mathcal{S}}$ is the approximate set, then the *accuracy* of the approximate algorithm can be reported as $|\mathcal{S} \cap \tilde{\mathcal{S}}|/k$.[3]

This monograph primarily studies the approximate[4] retrieval problem. As such, while we state a retrieval problem using the arg max or arg min notation, we are generally only interested in approximate solutions to it.

---

[3] This quantity is also known as *recall* in the literature, because we are counting the number of vectors our algorithm recalls from the exact solution set.

[4] We drop ϵ from the name when it is clear from context.

References

Y. Bai, X. Li, G. Wang, C. Zhang, L. Shang, J. Xu, Z. Wang, F. Wang, and Q. Liu. Sparterm: Learning term-based sparse representation for fast text retrieval, 2020.

S. Bruch, S. Gai, and A. Ingber. An analysis of fusion functions for hybrid retrieval. *ACM Transactions on Information Systems*, 42(1), 8 2023.

T. Chen, M. Zhang, J. Lu, M. Bendersky, and M. Najork. Out-of-domain semantics to the rescue! zero-shot hybrid retrieval models. In *Advances in Information Retrieval: 44th European Conference on IR Research*, pages 95–110, 2022.

Z. Dai and J. Callan. Context-aware term weighting for first stage passage retrieval. In *Proceedings of the 43rd International ACM SIGIR Conference on Research and Development in Information Retrieval*, pages 1533–1536, 2020.

T. Formal, B. Piwowarski, and S. Clinchant. Splade: Sparse lexical and expansion model for first stage ranking. In *Proceedings of the 44th International ACM SIGIR Conference on Research and Development in Information Retrieval*, pages 2288–2292, 2021.

T. Formal, C. Lassance, B. Piwowarski, and S. Clinchant. From distillation to hard negative sampling: Making sparse neural ir models more effective. In *Proceedings of the 45th International ACM SIGIR Conference on Research and Development in Information Retrieval*, page 2353–2359, 2022.

L. Gao, Z. Dai, and J. Callan. COIL: revisit exact lexical match in information retrieval with contextualized inverted list. In *Proceedings of the 2021 Conference of the North American Chapter of the Association for Computational Linguistics: Human Language Technologies, NAACL-HLT 2021, Online, June 6-11, 2021*, pages 3030–3042, 2021.

W. Guo, J. Wang, and S. Wang. Deep multimodal representation learning: A survey. *IEEE Access*, 7:63373–63394, 2019.

V. Karpukhin, B. Oguz, S. Min, P. Lewis, L. Wu, S. Edunov, D. Chen, and W.-t. Yih. Dense passage retrieval for open-domain question answering. In *Proceedings of the 2020 Conference on Empirical Methods in Natural Language Processing*, pages 6769–6781, Nov. 2020.

O. Khattab and M. Zaharia. Colbert: Efficient and effective passage search via contextualized late interaction over bert. In *Proceedings of the 43rd International ACM SIGIR Conference on Research and Development in Information Retrieval*, pages 39–48, 2020.

S. Kuzi, M. Zhang, C. Li, M. Bendersky, and M. Najork. Leveraging semantic and lexical matching to improve the recall of document retrieval systems: A hybrid approach, 2020.

J. Lin and X. Ma. A few brief notes on deepimpact, coil, and a conceptual framework for information retrieval techniques, 2021.

J. Lin, R. Nogueira, and A. Yates. *Pretrained Transformers for Text Ranking: BERT and Beyond.* Springer Cham, 2021.

J. Ma, I. Korotkov, K. Hall, and R. T. McDonald. Hybrid first-stage retrieval models for biomedical literature. In *CLEF*, 2020.

X. Ma, K. Sun, R. Pradeep, and J. J. Lin. A replication study of dense passage retriever, 2021.

A. Mallia, O. Khattab, T. Suel, and N. Tonellotto. Learning passage impacts for inverted indexes. In *Proceedings of the 44th International ACM SIGIR Conference on Research and Development in Information Retrieval*, pages 1723–1727, 2021.

N. Reimers and I. Gurevych. Sentence-bert: Sentence embeddings using siamese bert-networks. In *Proceedings of the 2019 Conference on Empirical Methods in Natural Language Processing*. Association for Computational Linguistics, 11 2019.

G. Salton and C. Buckley. Term-weighting approaches in automatic text retrieval. *Information Processing and Management*, 24(5):513–523, 1988.

K. Santhanam, O. Khattab, J. Saad-Falcon, C. Potts, and M. Zaharia. ColBERTv2: Effective and efficient retrieval via lightweight late interaction. In *Proceedings of the 2022 Conference of the North American Chapter of the Association for Computational Linguistics: Human Language Technologies*, pages 3715–3734, July 2022.

A. Vaswani, N. Shazeer, N. Parmar, J. Uszkoreit, L. Jones, A. N. Gomez, L. Kaiser, and I. Polosukhin. Attention is all you need. In *Proceedings of the 31st International Conference on Neural Information Processing Systems*, page 6000–6010, 2017.

S. Wang, S. Zhuang, and G. Zuccon. Bert-based dense retrievers require interpolation with bm25 for effective passage retrieval. In *Proceedings of the 2021 ACM SIGIR International Conference on Theory of Information Retrieval*, page 317–324, 2021.

X. Wu, R. Guo, D. Simcha, D. Dopson, and S. Kumar. Efficient inner product approximation in hybrid spaces, 2019.

L. Xiong, C. Xiong, Y. Li, K.-F. Tang, J. Liu, P. Bennett, J. Ahmed, and A. Overwijk. Approximate nearest neighbor negative contrastive learning for dense text retrieval. In *International Conference on Learning Representations*, 4 2021.

H. Zamani, M. Dehghani, W. B. Croft, E. Learned-Miller, and J. Kamps. From neural re-ranking to neural ranking: Learning a sparse representation for inverted indexing. In *Proceedings of the 27th ACM International Conference on Information and Knowledge Management*, pages 497–506, 2018.

C. Zhang, Z. Yang, X. He, and L. Deng. Multimodal intelligence: Representation learning, information fusion, and applications. *IEEE Journal of Selected Topics in Signal Processing*, 14(3):478–493, 2020.

S. Zhuang and G. Zuccon. Fast passage re-ranking with contextualized exact term matching and efficient passage expansion. In *Workshop on Reaching Efficiency in Neural Information Retrieval, the 45th International ACM*

SIGIR Conference on Research and Development in Information Retrieval, 2022.

Chapter 2
Retrieval Stability in High Dimensions

Abstract We are about to embark on a comprehensive survey and analysis of vector retrieval methods in the remainder of this monograph. It may thus sound odd to suggest that you may not need any of these clever ideas in order to perform vector retrieval. Sometimes, under bizarrely general conditions that we will explore formally in this chapter, an exhaustive search (where we compute the distance between query and every data point, sort, and return the top k) is likely to perform much better in both accuracy and search latency! The reason why that may be the case has to do with the approximate nature of algorithms and the oddities of high dimensions. We elaborate this point by focusing on the top-1 case.

2.1 Intuition

Consider the case of proper distance functions where $\delta(\cdot, \cdot)$ is a metric. Recall from Equation (1.5) that a vector u is an acceptable ϵ-approximate solution if its distance to the query q according to $\delta(\cdot, \cdot)$ is at most $(1 + \epsilon)\delta(q, u^*)$, where u^* is the optimal vector and ϵ is an arbitrary parameter. As shown in Figure 1.4(a) for NN, this means that, if you centered an L_p ball around q with radius $\delta(q, (1 + \epsilon)u^*)$, then u is in that ball.

So, what if we find ourselves in a situation where no matter how small ϵ is, too many vectors, or indeed *all* vectors, from our collection $\mathcal{X}$ end up in the $(1 + \epsilon)$-enlarged ball? Then, by definition, every vector is an ϵ-approximate nearest neighbor of q!

In such a configuration of points, it is questionable whether the notion of "nearest neighbor" has any meaning at all: If the query point were perturbed by some noise as small as ϵ, then its true nearest neighbor would suddenly change, making NN *unstable*. Because of that instability, any approximate algorithm will need to examine a large portion or nearly all of the data

S. Bruch, *Foundations of Vector Retrieval*, https://doi.org/10.1007/978-3-031-55182-6_2

points anyway, reducing thereby to a procedure that performs more poorly than exhaustive search.

That sounds troubling. But when might we experience that phenomenon? That is the question Beyer et al. [1999] investigate in their seminal paper.

It turns out, one scenario where vector retrieval becomes unstable as dimensionality d increases is if a) data points are *iid* in each dimension, b) query points are similarly drawn *iid* in each dimension, and c) query points are independent of data points. This includes many synthetic collections that are, even today, routinely but inappropriately used for evaluation purposes.

On the other hand, when data points form clusters and query points fall into these same clusters, then the (approximate) "nearest cluster" problem is meaningful—but not necessarily the approximate NN problem. So while it makes sense to use approximate algorithms to obtain the nearest cluster, search within clusters may as well be exhaustive. This, as we will learn in Chapter 7, is the basis for a popular and effective class of vector retrieval algorithms on real collections.

2.2 Formal Results

More generally, vector retrieval becomes unstable in high dimensions when the variance of the distance between query and data points grows substantially more slowly than its expected value. That makes sense. Intuitively, that means that more and more data points fall into the $(1+\epsilon)$-enlarged ball centered at the query. This can be stated formally as the following theorem due to Beyer et al. [1999], extended to any general distance function $\delta(\cdot, \cdot)$.

Theorem 2.1 *Suppose m data points $\mathcal{X} \subset \mathbb{R}^d$ are drawn iid from a data distribution and a query point q is drawn independent of data points from any distribution. Denote by X a random data point. If*

$$\lim_{d \to \infty} \operatorname{Var}\left[\delta(q, X)\right] / \, \mathbb{E}\left[\delta(q, X)\right]^2 = 0,$$

then for any $\epsilon > 0$, $\lim_{d \to \infty} \mathbb{P}\left[\delta(q, X) \leq (1+\epsilon)\delta(q, u^*)\right] = 1$, where u^* is the vector closest to q.

Proof. Let $\delta_* = \max_{u \in \mathcal{X}} \delta(q, u)$ and $\delta^* = \min_{u \in \mathcal{X}} \delta(q, u)$. If we could show that, for some d-dependent *positive* α and β such that $\beta/\alpha = 1 + \epsilon$, $\lim_{d \to \infty} \mathbb{P}\left[\alpha \leq \delta^* \leq \delta_* \leq \beta\right] = 1$, then we are done. That is because, in that case $\delta_*/\delta^* \leq \beta/\alpha = 1 + \epsilon$ almost surely and the claim follows.

From the above, all that we need to do is to find α and β for a given d. Intuitively, we want the interval $[\alpha, \beta]$ to contain $\mathbb{E}[\delta(q, X)]$, because we know from the condition of the theorem that the distances should concentrate around their mean. So $\alpha = (1-\eta)\,\mathbb{E}[\delta(q, X)]$ and $\beta = (1+\eta)\,\mathbb{E}[\delta(q, X)]$ for some η seems like a reasonable choice. Letting $\eta = \epsilon/(\epsilon+2)$ gives us the desired ratio: $\beta/\alpha = 1+\epsilon$.

Now we must show that δ_* and δ^* belong to our chosen $[\alpha, \beta]$ interval almost surely in the limit. That happens if *all* distances belong to that interval. So:

$$
\begin{aligned}
&\lim_{d\to\infty} \mathbb{P}\left[\alpha \leq \delta^* \leq \delta_* \leq \beta\right] = \\
&\quad \lim_{d\to\infty} \mathbb{P}\left[\delta(q,u) \in [\alpha, \beta] \quad \forall\, u \in \mathcal{X}\right] = \\
&\quad \lim_{d\to\infty} \mathbb{P}\left[(1-\eta)\,\mathbb{E}[\delta(q,X)] \leq \delta(q,u) \leq (1+\eta)\,\mathbb{E}[\delta(q,X)]\ \forall\, u \in \mathcal{X}\right] = \\
&\quad \lim_{d\to\infty} \mathbb{P}\left[\left|\delta(q,u) - \mathbb{E}[\delta(q,X)]\right| \leq \eta\,\mathbb{E}[\delta(q,X)] \quad \forall\, u \in \mathcal{X}\right].
\end{aligned}
$$

It is now easier to work with the complementary event:

$$
1 - \lim_{d\to\infty} \mathbb{P}\left[\exists\, u \in \mathcal{X}\ s.t.\quad \left|\delta(q,u) - \mathbb{E}[\delta(q,X)]\right| > \eta\,\mathbb{E}[\delta(q,X)]\right].
$$

Using the Union Bound, the probability above is greater than or equal to the following:

$$
\begin{aligned}
&\lim_{d\to\infty} \mathbb{P}\left[\alpha \leq \delta^* \leq \delta_* \leq \beta\right] \geq \\
&\quad 1 - \lim_{d\to\infty} \sum_{u\in\mathcal{X}} \mathbb{P}\left[\left|\delta(q,u) - \mathbb{E}[\delta(q,X)]\right| > \eta\,\mathbb{E}[\delta(q,X)]\right] = \\
&\quad 1 - \lim_{d\to\infty} \sum_{u\in\mathcal{X}} \mathbb{P}\left[\left(\delta(q,u) - \mathbb{E}[\delta(q,X)]\right)^2 > \eta^2\,\mathbb{E}[\delta(q,X)]^2\right].
\end{aligned}
$$

Note that, q is independent of data points and that data points are *iid* random variables. Therefore, $\delta(q,u)$'s are random variables drawn *iid* as well. Furthermore, by assumption $\mathbb{E}[\delta(q,X)]$ exists, making it possible to apply Markov's inequality to obtain the following bound:

$$
\begin{aligned}
&\lim_{d\to\infty} \mathbb{P}\left[\alpha \leq \delta^* \leq \delta_* \leq \beta\right] \geq \\
&\quad 1 - \lim_{d\to\infty} |\mathcal{X}|\,\mathbb{P}\left[\left(\delta(q,X) - \mathbb{E}[\delta(q,X)]\right)^2 > \eta^2\,\mathbb{E}[\delta(q,X)]^2\right] \geq \\
&\quad 1 - \lim_{d\to\infty} m \frac{1}{\eta^2\,\mathbb{E}\left[\delta(q,X)\right]^2}\,\mathbb{E}\left[\left(\delta(q,u) - \mathbb{E}[\delta(q,X)]\right)^2\right] = \\
&\quad 1 - \lim_{d\to\infty} m \frac{\mathrm{Var}[\delta(q,X)]}{\eta^2\,\mathbb{E}[\delta(q,X)]^2}.
\end{aligned}
$$

By the conditions of the theorem, $\text{Var}[\delta(q, X)]/\,\mathbb{E}[\delta(q, X)]^2 \rightarrow 0$ as $d \rightarrow \infty$, so that the last expression tends to 1 in the limit. That concludes the proof. □

We mentioned earlier that if data and query points are independent of each other and that vectors are drawn *iid* in each dimension, then vector retrieval becomes unstable. For NN with the L_p norm, it is easy to show that such a configuration satisfies the conditions of Theorem 2.1, hence the instability. Consider the following for $\delta(q, u) = \|q - u\|_p^p$:

$$\lim_{d\to\infty} \frac{\text{Var}\left[\|q-u\|_p^p\right]}{\mathbb{E}\left[\|q-u\|_p^p\right]^2} = \lim_{d\to\infty} \frac{\text{Var}\left[\sum_i (q_i - u_i)^p\right]}{\mathbb{E}\left[\sum_i (q_i - u_i)^p\right]^2} =$$

$$\lim_{d\to\infty} \frac{\sum_i \text{Var}\left[(q_i - u_i)^p\right]}{\left(\sum_i \mathbb{E}\left[(q_i - u_i)^p\right]\right)^2} \qquad \text{(by independence)}$$

$$\lim_{d\to\infty} \frac{d\sigma^2}{d^2\mu^2} = 0,$$

where we write $\sigma^2 = \text{Var}[(q_i - u_i)^p]$ and $\mu = \mathbb{E}[(q_i - u_i)^p]$.

When $\delta(q, u) = -\langle q, u \rangle$, the same conditions result in retrieval instability:

$$\lim_{d\to\infty} \frac{\text{Var}\left[\langle q, u\rangle\right]}{\mathbb{E}\left[\langle q, u\rangle\right]^2} = \lim_{d\to\infty} \frac{\text{Var}\left[\sum_i q_i u_i\right]}{\mathbb{E}\left[\sum_i q_i u_i\right]^2} =$$

$$\lim_{d\to\infty} \frac{\sum_i \text{Var}\left[q_i u_i\right]}{\left(\sum_i \mathbb{E}\left[q_i u_i\right]\right)^2} \qquad \text{(by independence)}$$

$$\lim_{d\to\infty} \frac{d\sigma^2}{d^2\mu^2} = 0,$$

where we write $\sigma^2 = \text{Var}[q_i u_i]$ and $\mu = \mathbb{E}[q_i u_i]$.

2.3 Empirical Demonstration of Instability

Let us examine the theorem empirically. We simulate the NN setting with L_2 distance and report the results in Figure 2.1. In these experiments, we sample 1,000,000 data points with each coordinate drawing its value independently from the same distribution, and 1,000 query points sampled similarly. We then compute the minimum and maximum distance between each query point and the data collection, measure the ratio between them, and report the mean and standard deviation of the ratio across queries. We repeat this exercise for various values of dimensionality d and render the results in Figure 2.1(a). Unsurprisingly, this ratio tends to 1 as $d \rightarrow \infty$, as predicted by the theorem.

Another way to understand this result is to count the number of data points that qualify as approximate nearest neighbors. The theory predicts

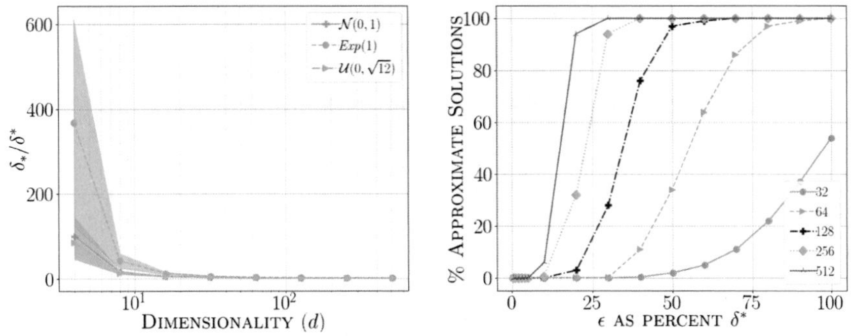

(a) δ_*/δ^* (b) PERCENT APPROXIMATE SOLUTIONS

Fig. 2.1: Simulation results for Theorem 2.1 applied to NN with L_2 distance. *Left*: The ratio between the maximum distance between a query and data points δ_*, to the minimum distance δ^*. The shaded region shows one standard deviation. As dimensionality increases, this ratio tends to 1. *Right*: The percentage of data points whose distance to a query is at most $(1+\epsilon/100)\delta^*$, visualized for the Gaussian distribution—the trend is similar for other distributions. As d increases, more vectors fall into the enlarged ball, making them valid solutions to the approximate NN problem.

that, as d increases, we can find a smaller ϵ such that nearly all data points fall within $(1+\epsilon)\delta^*$ distance from the query. The results of our experiments confirm this phenomenon; we have plotted the results for the Gaussian distribution in Figure 2.1(b).

2.3.1 Maximum Inner Product Search

In the discussion above, we established that retrieval becomes unstable in high dimensions if the data satisfies certain statistical conditions. That meant that the difference between the maximum and the minimum distance grows just as fast as the magnitude of the minimum distance, so that any approximate solution becomes meaningless.

> The instability statement does not necessarily imply, however, that the distances become small or converge to a certain value. But as we see in this section, inner product in high dimensions does become smaller and smaller as a function of d.

The following theorem summarizes this phenomenon for a unit query point and bounded data points. Note that, the condition that q is a unit vector is not restrictive in any way, as the norm of the query point does not affect the retrieval outcome.

Theorem 2.2 *If m data points with bounded norms, and a unit query vector q are drawn* iid *from a spherically symmetric*[1] *distribution in $\mathbb{R}^d$, then:*

$$\lim_{d \to \infty} \mathbb{P}\left[\langle q, X \rangle > \epsilon\right] = 0.$$

Proof. By spherical symmetry, it is easy to see that $\mathbb{E}[\langle q, X \rangle] = 0$. The variance of the inner product is then equal to $\mathbb{E}[\langle q, X \rangle^2]$, which can be expanded as follows.

First, find an orthogonal transformation $\Gamma : \mathbb{R}^d \to \mathbb{R}^d$ that maps the query point q to the first standard basis (i.e., $e_1 = [1, 0, 0, \ldots, 0] \in \mathbb{R}^d$). Due to spherical symmetry, this transformation does not change the data distribution. Now, we can write:

$$\mathbb{E}[\langle q, X \rangle^2] = \mathbb{E}[\langle \Gamma q, \Gamma X \rangle^2] = \mathbb{E}[(\Gamma X)_1^2] =$$

$$\mathbb{E}[\frac{1}{d}\sum_{i=1}^{d}(\Gamma X)_i^2] = \frac{1}{d}\|X\|_2^2.$$

In the above, the third equality is due to the fact that the distribution of the (transformed) vectors is the same in every direction. Because $\|X\|$ is bounded by assumption, the variance of the inner product between q and a random data point tends to 0 as $d \to \infty$. The claim follows. □

The proof of Theorem 2.2 tells us that the variance of inner product grows as a function of $1/d$ and $\|X\|_2^2$. So if our vectors have bounded norms, then we can find a d such that inner products are arbitrarily close to 0. This is yet another reason that approximate MIPS could become meaningless. But if our data points are clustered in (near) orthogonal subspaces, then approximate MIPS over clusters makes sense—though, again, MIPS within clusters would be unstable.

References

K. Beyer, J. Goldstein, R. Ramakrishnan, and U. Shaft. When is "nearest neighbor" meaningful? In *Database Theory*, pages 217–235, 1999.

[1] A distribution is spherically symmetric if it remains invariant under an orthogonal transformation.

Chapter 3
Intrinsic Dimensionality

Abstract We have seen that high dimensionality poses difficulties for vector retrieval. Yet, judging by the progression from hand-crafted feature vectors to sophisticated embeddings of data, we detect a clear trend towards higher dimensional representations of data. How worried should we be about this ever increasing dimensionality? This chapter explores that question. Its key message is that, even though data points may appear to belong to a high-dimensional space, they actually lie on or near a low-dimensional manifold and, as such, have a low *intrinsic* dimensionality. This chapter then formalizes the notion of intrinsic dimensionality and presents a mathematical framework that will be useful in analyses in future chapters.

3.1 High-Dimensional Data and Low-Dimensional Manifolds

We talked a lot about the difficulties of answering ϵ-approximate top-k questions in high dimensions. We said, in certain situations, the question itself becomes meaningless and retrieval falls apart. For MIPS, in particular, we argued in Theorem 2.2 that points become nearly orthogonal almost surely as the number of dimensions increases. But how concerned should we be, especially given the ever-increasing dimensionality of vector representations of data? Do our data points really live in such extremely high-dimensional spaces? Are all the dimensions necessary to preserving the structure of our data or do our data points have an intrinsically smaller dimensionality?

The answer to these questions is sometimes obvious. If a set of points in $\mathbb{R}^d$ lie strictly in a flat subspace $\mathbb{R}^{d_\circ}$ with $d_\circ < d$, then one can simply drop the "unused" dimensions—perhaps after a rotation. This could happen if a pair of coordinates are correlated, for instance. No matter what query vector we are performing retrieval for or what distance function we use, the top-k

S. Bruch, *Foundations of Vector Retrieval*, https://doi.org/10.1007/978-3-031-55182-6_3

set does not change whether the unused dimensions are taken into account or the vectors corrected to lie in $\mathbb{R}^{d_\circ}$.

Other times the answer is intuitive but not so obvious. When a text document is represented as a sparse vector, all the document's information is contained entirely in the vector's non-zero coordinates. The coordinates that are 0 do not contribute to the representation of the document in any way. In a sense then, the intrinsic dimensionality of a collection of such sparse vectors is in the order of the number of non-zero coordinates, rather than the nominal dimensionality of the space the points lie in.

It appears then that there are instances where a collection of points have a superficially large number of dimensions, d, but that, in fact, the points lie in a lower-dimensional space with dimensionality $d_\circ$. We call $d_\circ$ the *intrinsic dimensionality* of the point set.

This situation, where the intrinsic dimensionality of data is lower than that of the space, arises more commonly than one imagines. In fact, so common is this phenomenon that in statistical learning theory, there are special classes of algorithms [Ma and Fu, 2012] designed for data collections that lie on or near a low-dimensional submanifold of $\mathbb{R}^d$ despite their apparent arbitrarily high-dimensional representations.

In the context of vector retrieval, too, the concept of intrinsic dimensionality often plays an important role. Knowing that data points have a low intrinsic dimensionality means we may be able to reduce dimensionality without (substantially) losing the geometric structure of the data, including interpoint distances. But more importantly, we can design algorithms specifically for data with low intrinsic dimensionality, as we will see in later chapters. In our analysis of many of these algorithms, too, we often resort to this property to derive meaningful bounds and make assertions about their performance.

Doing so, however, requires that we formalize the notion of intrinsic dimensionality. We often do not have a characterization of the submanifold itself, so we need an alternate way of characterizing the low-dimensional structure of our data points. In the remainder of this chapter, we present two common (and related) definitions of intrinsic dimensionality that will be useful in subsequent chapters.

3.2 Doubling Measure and Expansion Rate

Karger and Ruhl [2002] characterize intrinsic dimensionality as the growth or *expansion* rate of a point set. To understand what that means intuitively, place yourself somewhere in the data collection, draw a ball around yourself, and count how many data points are in that ball. Now expand the radius of this ball by a factor 2, and count again. The count of data points in a "growth-restricted" point set should increase *smoothly*, rather than suddenly, as we make this ball larger.

In other words, data points "come into view," as Karger and Ruhl [2002] put it, at a constant rate as we expand our view, regardless of where we are located. We will not encounter massive holes in the space where there are no data points, followed abruptly by a region where a large number of vectors are concentrated.

The formal definition is not far from the intuitive description above. In fact, expansion rate as defined by Karger and Ruhl [2002] is an instance of the following more general definition of a *doubling measure*, where the measure μ is the counting measure over a collection of points $\mathcal{X}$.

Definition 3.1 A distribution μ on $\mathbb{R}^d$ is a *doubling measure* if there is a constant $d_\circ$ such that, for any $r > 0$ and $x \in \mathbb{R}^d$, $\mu(B(x, 2r)) \leq 2^{d_\circ}\mu(B(x, r))$. The constant $d_\circ$ is said to be the *expansion rate* of the distribution.

One can think of the expansion rate $d_\circ$ as a dimension of sorts. In fact, as we will see later, several works [Dasgupta and Sinha, 2015, Karger and Ruhl, 2002, Beygelzimer et al., 2006] use this notion of intrinsic dimensionality to design algorithms for top-k retrieval or utilize it to derive performance guarantees for vector collections that are drawn from a doubling measure. That is the main reason we review this definition of intrinsic dimensionality in this chapter.

While the expansion rate is a reasonable way of describing the structure of a set of points, it is unfortunately not a stable indicator. It can suddenly blow up, for example, by the addition of a single point to the set. As a concrete example, consider the set of integers between $|r|$ and $|2r|$ for any arbitrary value of r: $\mathcal{X} = \{u \in \mathbb{Z} \mid r < |u| < 2r\}$. The expansion rate of the resulting set is constant because no matter which point we choose as the center of our ball, and regardless of our choice of radius, doubling the radius brings points into view at a constant rate.

What happens if we added the origin to the set, so that our set becomes $\{0\} \cup \mathcal{X}$? If we chose 0 as the center of the ball, and set its radius to r, we have a single point in the resulting ball. The moment we double r, the resulting ball will contain the entire set! In other words, the expansion rate of the updated set is $\log m$ (where $m = |\mathcal{X}|$).

It is easy to argue that a subset of a set with bounded expansion rate does not necessarily have a bounded expansion rate itself. This unstable behavior is less than ideal, which is why a more robust notion of intrinsic dimensionality has been developed. We will introduce that next.

3.3 Doubling Dimension

Another idea to formalize intrinsic dimensionality that has worked well in algorithmic design and anlysis is the *doubling dimension.* It was introduced by Gupta et al. [2003] but is closely related to the Assouad dimension [Assouad, 1983]. It is defined as follows.

Definition 3.2 A set $\mathcal{X} \subset \mathbb{R}^d$ is said to have doubling dimension $d_\circ$ if $B(\cdot, 2r) \cap \mathcal{X}$, the intersection of any ball of radius $2r$ with the set, can be covered by at most $2^{d_\circ}$ balls of radius r.

The base 2 in the definition above can be replaced with any other constant k: The doubling dimension of $\mathcal{X}$ is $d_\circ$ if the intersection of any ball of radius r with the set can be covered by $\mathcal{O}(k^{d_\circ})$ balls of radius r/k. Furthermore, the definition can be easily extended to any metric space, not just $\mathbb{R}^d$ with the Euclidean norm.

The doubling dimension is a different notion from the expansion rate as defined in Definition 3.1. The two, however, are in some sense related, as the following lemma shows.

Lemma 3.1 *The doubling dimension,* $d_\circ$ *of any finite metric* (X, δ) *is bounded above by its expansion rate,* $d_\circ^{\text{KR}}$ *times 4:* $d_\circ \leq 4d_\circ^{\text{KR}}$.

Proof. Fix a ball $B(u, 2r)$ and let S be its r-net. That is, $S \subset X$, the distance between any two points in S is at least r, and $\mathcal{X} \subseteq \bigcup_{u \in S} B(u, r)$. We have that:

$$B(u, 2r) \subset \bigcup_{v \in S} B(v, r) \subset B(u, 4r).$$

By definition of the expansion rate, for every $v \in S$:

$$\lvert B(u, 4r) \rvert \leq \lvert B(v, 8r) \rvert \leq 2^{4d_\circ^{\text{KR}}} \lvert B(v, \frac{r}{2}) \rvert.$$

Because the balls $B(v, r/2)$ for all $v \in S$ are disjoint, it follows that $|S| \leq 2^{4d_\circ^{\text{KR}}}$, so that $2^{4d_\circ^{\text{KR}}}$ many balls of radius r cover $B(u, 2r)$. That concludes the proof. □

The doubling dimension and expansion rate both quantify the intrinsic dimensionality of a point set. But Lemma 3.1 shows that, the class of doubling metrics (i.e., metric spaces with a constant doubling dimension) contains the class of metrics with a bounded expansion rate.

The converse of the above lemma is not true. In other words, there are sets that have a bounded doubling dimension, but whose expansion rate is unbounded. The set, $\mathcal{X} = \{0\} \cup \{u \in \mathbb{Z} \mid r < |u| < 2r\}$, from the previous section is one example where this happens. From our discussion above, its expansion rate is $\log|\mathcal{X}|$. It is easy to see that the doubling dimension of this set, however, is constant.

3.3.1 Properties of the Doubling Dimension

It is helpful to go over a few concrete examples of point sets with bounded doubling dimension in order to understand a few properties of this definition of intrinsic dimensionality. We will start with a simple example: a line segment in $\mathbb{R}^d$ with the Euclidean norm.

If the set $\mathcal{X}$ is a line segment, then its intersection with a ball of radius r is itself a line segment. Clearly, the intersection set can be covered with two balls of radius $r/2$. Therefore, the doubling dimension $d_\circ$ of $\mathcal{X}$ is 1.

We can extend that result to any affine set in $\mathbb{R}^d$ to obtain the following property:

Lemma 3.2 *A k-dimensional flat in $\mathbb{R}^d$ has doubling dimension $\mathcal{O}(k)$.*

Proof. The intersection of a ball in $\mathbb{R}^d$ and a k-dimensional flat is a ball in $\mathbb{R}^k$. It is a well-known result that the size of an ϵ-net of a unit ball in $\mathbb{R}^k$ is at most $(C/\epsilon)^k$ for some small constant C. As such, a ball of radius r can be covered with $2^{\mathcal{O}(k)}$ balls of radius $r/2$, implying the claim. □

The lemma above tells us that the doubling dimension of a set in the Euclidean space is at most some constant factor larger than the natural dimension of the space; note that this was not the case for the expansion rate. Another important property that speaks to the stability of the doubling dimension is the following, which is trivially true:

Lemma 3.3 *Any subset of a set with doubling dimension $d_\circ$ itself has doubling dimension $d_\circ$.*

The doubling dimension is also robust under the addition of points to the set, as the following result shows.

Lemma 3.4 *Suppose sets $\mathcal{X}_i$ for $i \in [n]$ each have doubling dimension $d_\circ$. Then their union has doubling dimension at most $d_\circ + \log n$.*

Proof. For any ball B of radius r, $B \cap \mathcal{X}_i$ can be covered with $2^{d_\circ}$ balls of half the radius. As such, at most $n2^{d_\circ}$ balls of radius $r/2$ are needed to cover the union. The doubling dimension of the union is therefore $d_\circ + \log n$. □

One consequence of the previous two lemmas is the following statement concerning sparse vectors:

Lemma 3.5 *Suppose that $\mathcal{X} \subset \mathbb{R}^d$ is a collection of sparse vectors, each having at most n non-zero coordinates. Then the doubling dimension of $\mathcal{X}$ is at most $Ck + k \log d$ for some constant C.*

Proof. $\mathcal{X}$ is the union of $\binom{d}{n} \leq d^n$ n-dimensional flats. Each of these flats has doubling dimension Ck for some universal constant C, by Lemma 3.2. By the application of Lemma 3.4, we get that the doubling dimension of $\mathcal{X}$ is at most $Cn + n \log d$. □

> Lemma 3.5 states that collections of sparse vectors in the Euclidean space are naturally described by the doubling dimension.

References

P. Assouad. Plongements lipschitziens dans $\mathbb{R}^n$. *Bulletin de la Société Mathématique de France*, 111:429–448, 1983.

A. Beygelzimer, S. Kakade, and J. Langford. Cover trees for nearest neighbor. In *Proceedings of the 23rd International Conference on Machine Learning*, page 97–104, 2006.

S. Dasgupta and K. Sinha. Randomized partition trees for nearest neighbor search. *Algorithmica*, 72(1):237–263, 5 2015.

A. Gupta, R. Krauthgamer, and J. Lee. Bounded geometries, fractals, and low-distortion embeddings. In *44th Annual IEEE Symposium on Foundations of Computer Science*, pages 534–543, 2003.

D. R. Karger and M. Ruhl. Finding nearest neighbors in growth-restricted metrics. In *Proceedings of the 34th Annual ACM Symposium on Theory of Computing*, pages 741–750, 2002.

Y. Ma and Y. Fu. *Manifold Learning Theory and Applications.* CRC Press, 2012.

Part II
Retrieval Algorithms

Chapter 4
Branch-and-Bound Algorithms

Abstract One of the earliest approaches to the top-k retrieval problem is to partition the vector space recursively into smaller regions and, each time we do so, make note of their geometry. During search, we eliminate the regions whose shape indicates they cannot contain or overlap with the solution set. This chapter covers algorithms that embody this approach and discusses their exact and approximate variants.

4.1 Intuition

Suppose there was some way to split a collection $\mathcal{X}$ into two sub-collections, $\mathcal{X}_l$ and $\mathcal{X}_r$, such that $\mathcal{X} = \mathcal{X}_l \cup \mathcal{X}_r$ and that the two sub-collections have roughly the same size. In general, we can relax the splitting criterion so the two sub-collections are not necessarily partitions; that is, we may have $\mathcal{X}_l \cap \mathcal{X}_r \neq \emptyset$. We may also split the collection into more than two sub-collections. For the moment, though, assume we have two sub-collections that do not overlap.

Suppose further that, we could geometrically characterize *exactly* the regions that contain $\mathcal{X}_l$ and $\mathcal{X}_r$. For example, when $\mathcal{X}_l \cap \mathcal{X}_r = \emptyset$, these regions partition the space and may therefore be characterized by a separating hyperplane. Call these regions $\mathcal{R}_l$ and $\mathcal{R}_r$, respectively. The separating hyperplane forms a *decision boundary* that helps us determine if a vector falls into $\mathcal{R}_l$ or $\mathcal{R}_r$.

In effect, we have created a binary tree of depth 1 where the root node has a decision boundary and each of the two leaves contains data points that fall into its region. This is illustrated in Figure 4.1(a).

Now suppose we have a query point q somewhere in the space and that we are interested in finding the top-1 data point with respect to a proper distance function $\delta(\cdot, \cdot)$. q falls either in $\mathcal{R}_l$ or $\mathcal{R}_r$; suppose it is in $\mathcal{R}_l$. We determine that by evaluating the decision boundary in the root of the tree

S. Bruch, *Foundations of Vector Retrieval*, https://doi.org/10.1007/978-3-031-55182-6_4

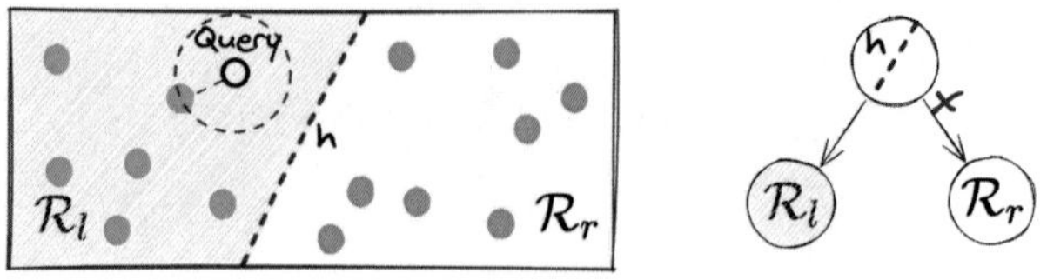

(a)

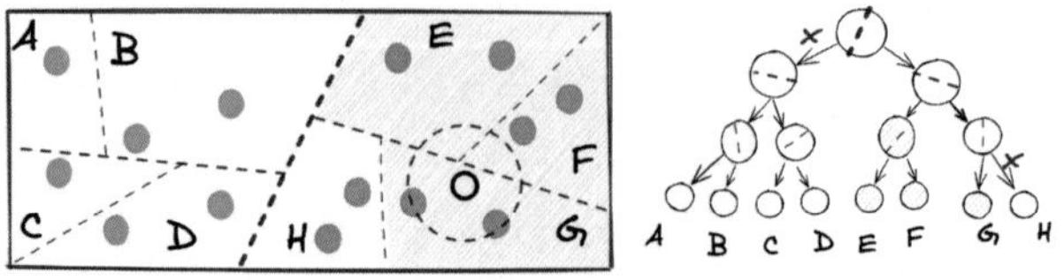

(b)

Fig. 4.1: Illustration of a general branch-and-bound method on a toy collection in $\mathbb{R}^2$. In (a), $\mathcal{R}_l$ and $\mathcal{R}_r$ are separated by the dashed line h. The distance between query q and the closest vector in $\mathcal{R}_l$ is less than the distance between q and h. As such, we do not need to search for the top-1 vector over the points in $\mathcal{R}_r$, so that the right branch of the tree is pruned. In (b), the regions are recursively split until each terminal region contains at most two data points. We then find the distance between q and the data points in the region that contains q, G. If the ball around q with this distance as its radius does not intersect a region, we can safely prune that region—regions that are not shaded in the figure. Otherwise, we may have to search it during the certification process.

and navigating to the appropriate leaf. Now, we solve the exact top-1 retrieval problem over $\mathcal{X}_l$ to obtain the optimal point in that region u_l^*, then make a note of this minimum distance obtained, $\delta(q, u_l^*)$.

At this point, if it turns out that $\delta(q, u_l^*) < \delta(q, \mathcal{R}_r)$[1] then we have found the optimal point and do not need to search the data points in $\mathcal{X}_r$ at all! That is because, the δ-ball[2] centered at q with radius $\delta(q, u_l^*)$ is contained entirely in $\mathcal{R}_l$, so that no point from $\mathcal{R}_r$ can have a shorter distance to q than u_l^*. Refer again to Figure 4.1(a) for an illustration of this scenario.

If, on the other hand, $\delta(q, u_l^*) \geq \delta(q, \mathcal{R}_r)$, then we proceed to solve the top-1 problem over $\mathcal{X}_r$ as well and compare the solution with u_l^* to find the optimal vector. We can think of the comparison between $\delta(q, u_l^*)$ with $\delta(q, \mathcal{R}_r)$ as backtracking to the parent node of $\mathcal{R}_l$ in the equivalent tree—which is the root—and comparing $\delta(q, u_l^*)$ with the distance of q with the decision boundary. This process of backtracking and deciding to prune a

[1] The distance between a point u and a set $\mathcal{S}$ is defined as $\delta(u, \mathcal{S}) = \inf_{v \in \mathcal{S}} \delta(u, v)$.

[2] The ball centered at u with radius r with respect to metric δ is $\{x \mid \delta(u, x) \leq r\}$.

branch or search it *certifies* that u_l^* is indeed optimal, thereby solving the top-1 problem exactly.

We can extend the framework above easily by recursively splitting the two sub-collections and characterizing the regions containing the resulting partitions. This leads to a (balanced) binary tree where each internal node has a decision boundary—the separating hyperplane of its child regions. We may stop splitting a node if it has fewer than $m_\circ$ points. This extension is rendered in Figure 4.1(b).

The retrieval process is the same but needs a little more care: Let the query q traverse the tree from root to leaf, where each internal node determines if q belongs to the "left" or "right" sub-regions and *routes* q accordingly. Once we have found the leaf (terminal) region that contains q, we find the candidate vector u^*, then backtrack and certify that u^* is indeed optimal.

During the backtracking, at each internal node, we compare the distance between q and the current candidate with the distance between q and the region on the other side of the decision boundary. As before, that comparison results in either pruning a branch or searching it to find a possibly better candidate. The certification process stops when we find ourselves back in the root node with no more branches to verify, at which point we have found the optimal solution.

The above is the logic that is at the core of branch-and-bound algorithms for top-k retrieval [Dasgupta and Sinha, 2015, Bentley, 1975, Ram and Sinha, 2019, Ciaccia et al., 1997, Yianilos, 1993, Liu et al., 2004, Panigrahy, 2008, Ram and Gray, 2012, Bachrach et al., 2014]. The specific instances of this framework differ in terms of how they *split* a collection and the details of the *certification* process. We will review key algorithms that belong to this family in the remainder of this chapter. We emphasize that, most branch-and-bound algorithms only address the NN problem in the Euclidean space (so that $\delta(u, v) = \|u - v\|_2$) or in growth-restricted measures [Karger and Ruhl, 2002, Clarkson, 1997, Krauthgamer and Lee, 2004] but where the metric is nonetheless proper.

4.2 k-dimensional Trees

The k-dimensional Tree or k-d Tree [Bentley, 1975] is a special instance of the framework described above wherein the distance function is Euclidean and the space is recursively partitioned into hyper-rectangles. In other words, the decision boundaries in a k-d Tree are axis-aligned hyperplanes.

Let us consider its simplest construction for $\mathcal{X} \subset \mathbb{R}^d$. The root of the tree is a node that represents the entire space, which naturally contains the entire data collection. Assuming that the size of the collection is greater than 1, we follow a simple procedure to split the node: We select one coordinate axis and partition the collection at the median of data points along the chosen direction. The process recurses on each newly-minted node, with nodes at the same depth in the tree using the same coordinate axis for splitting, and where we go through the coordinates in a round-robin manner as the tree grows. We stop splitting a node further if it contains a single data point ($m_\circ = 1$), then mark it as a leaf node.

A few observations that are worth noting. By choosing the median point to split on, we guarantee that the tree is balanced. That together with the fact that $m_\circ = 1$ implies that the depth of the tree is $\log m$ where $m = |\mathcal{X}|$. Finally, because nodes in each level of the tree split on the same coordinate, every coordinate is split in $(\log m)/d$ levels. These will become important in our analysis of the algorithm.

4.2.1 Complexity Analysis

The k-d Tree data structure is fairly simple to construct. It is also efficient: Its space complexity given a set of m vectors is $\Theta(m)$ and its construction time has complexity $\Theta(m \log m)$.[3]

The search algorithm, however, is not so easy to analyze in general. Friedman et al. [1977] claimed that the expected search complexity is $\mathcal{O}(\log m)$ for m data points that are sampled uniformly from the unit hypercube. While uniformity is an unrealistic assumption, it is necessary for the analysis of the average case. On the other hand, no generality is lost by the assumption that vectors are contained in the hypercube. That is because, we can always scale every data point by a constant factor into the unit hypercube—a transformation that does not affect the pairwise distances between vectors. Let us now discuss the sketch of the proof of their claim.

Let $\delta^* = \min_{u \in \mathcal{X}} \|q - u\|_2$ be the optimal distance to a query q. Consider the ball of radius δ^* centered at q and denote it by $B(q, \delta^*)$. It is easy to see that the number of leaves we may need to visit in order to certify an initial candidate is upper-bounded by the number of leaf regions (i.e., d-dimensional boxes) that touch $B(q, \delta^*)$. That quantity itself is upper-bounded by the number of boxes that touch the smallest hypercube that contains $B(q, \delta^*)$. If we calculated this number, then we have found an upper-bound on the search complexity.

Following the argument above, Friedman et al. [1977] show that—with very specific assumptions on the density of vectors in the space, which do not necessarily hold in high dimensions—the quantity of interest is upper-bounded by the following expression:

$$\left(1 + G(d)^{1/d}\right)^d, \tag{4.1}$$

[3] The time to construct the tree depends on the complexity of the subroutine that finds the median of a set of values.

where $G(d)$ is the ratio between the volume of the hypercube that contains $B(q, \delta^*)$ and the volume of $B(q, \delta^*)$ itself. Because $G(d)$ is independent of m, and because visiting each leaf takes $\mathcal{O}(\log m)$ (i.e., the depth of the tree) operations, they conclude that the complexity of the algorithm is $\mathcal{O}(\log m)$.

4.2.2 Failure in High Dimensions

The argument above regarding the search time complexity of the algorithm fails in high dimensions. Let us elaborate why in this section.

Let us accept that the assumptions that enabled the proof above hold and focus on $G(d)$. The volume of a hypercube in d dimensions with sides that have length $2\delta^*$ is $(2\delta^*)^d$. The volume of $B(q, \delta^*)$ is $\pi^{d/2}{\delta^*}^d/\Gamma(d/2+1)$, where Γ denotes the Gamma function. For convenience, suppose that d is even, so that $\Gamma(d/2+1) = (d/2)!$. As such $G(d)$, the ratio between the two volumes, is:

$$G(d) = \frac{2^d (d/2)!}{\pi^{d/2}}. \tag{4.2}$$

Plugging this back into Equation (4.1), we arrive at:

$$\begin{aligned}
\left(1 + G(d)^{1/d}\right)^d &= \left(1 + \frac{2}{\sqrt{\pi}}(d/2)!^{\frac{1}{d}}\right)^d \\
&= \mathcal{O}\left(\left(\frac{2}{\sqrt{\pi}}\right)^d \left(\frac{d}{2}\right)!\right) \\
&= \mathcal{O}\left(\left(\frac{2}{\sqrt{\pi}}\right)^d d^{\frac{d+1}{2}}\right),
\end{aligned}$$

where in the third equality we used Stirling's formula, which approximates $n!$ as $\sqrt{2\pi n}(\frac{n}{e})^n$, to expand $(d/2)!$ as follows:

$$\begin{aligned}
\left(\frac{d}{2}\right)! &\approx \sqrt{2\pi d/2}\left(\frac{d}{2e}\right)^{d/2} \\
&= \sqrt{\pi}\frac{1}{(2e)^{d/2}} d^{\frac{d+1}{2}} \\
&= \mathcal{O}(d^{\frac{d+1}{2}}).
\end{aligned}$$

> The above shows that, the number of leaves that may be visited during the certification process has, asymptotically, an exponential dependency on d. That does not bode well for high dimensions.

There is an even simpler argument to make to show that in high dimensions the search algorithm must visit at least 2^d data points during the certification process. Our argument is as follows. We will show in Lemma 4.1 that, with high probability, the distance between the query point q and a randomly drawn data point concentrates sharply on $\sqrt{d}$. This implies that $B(q, \delta^*)$ has a radius that is larger than 1 with high probability. Noting that the side of the unit hypercube is 2, it follows that $B(q, \delta^*)$ crosses decision boundaries across every dimension, making it necessary to visit the corresponding partitions for certification.

Finally, because each level of the tree splits on a single dimension, the reasoning above means that the certification process must visit $\Omega(d)$ levels of the tree. As a result, we visit at least $2^{\Omega(d)}$ data points. Of course, in high dimensions, we often have far fewer than 2^d data points, so that we end up visiting every vector during certification.

Lemma 4.1 *The distance r between a randomly chosen point and its nearest neighbor among m points drawn uniformly at random from the unit hypercube is $\Theta\big(\sqrt{d}/m^{1/d}\big)$ with probability at least $1 - \mathcal{O}(1/2^d)$.*

Proof. Consider the ball of radius r in d-dimensional unit hypercube with volume 1. Suppose, for notational convenience, that d is even—whether it is odd or even does not change our asymptotic conclusions. The volume of this ball is:

$$\frac{\pi^{d/2} r^d}{(d/2)!}.$$

Since we have m points in the hypercube, the expected number of points that are contained in the ball of radius r is therefore:

$$\frac{\pi^{d/2} r^d}{(d/2)!} m.$$

As a result, the radius r for which the ball contains one point in expectation is:

$$\begin{aligned} \frac{\pi^{d/2} r^d}{(d/2)!} m = 1 \implies r^d &= \Theta\big(\frac{1}{m}(\frac{d}{2})!\big) \\ \implies r &= \Theta\big(\frac{1}{m^{1/d}}(\frac{d}{2})!^{1/d}\big). \end{aligned}$$

Using Stirling's formula and letting Θ consume the constants and small factors completes the claim that $r = \sqrt{d}/m^{1/d}$.

All that is left is bounding the probability of the event that r takes on the above value. For that, consider first the ball of radius $r/2$. The probability that this ball contains at least one point is at most $1/2^d$. To see this, note that the probability that a single point falls into this ball is:

$$\frac{\pi^{d/2}(r/2)^d}{(d/2)!} = \frac{1}{2^d} \underbrace{\frac{\pi^{d/2} r^d}{(d/2)!}}_{1/m}.$$

By the Union Bound, the probability that at least one point out of m points falls into this ball is at most $m \times 1/(m2^d) = 1/2^d$.

Next, consider the ball of radius $2r$. The probability that it contains no points at all is at most $(1 - 2^d/m)^m \approx \exp(-2^d) \leq 1/2^d$, where we used the approximation that $\left(1 - 1/x\right)^x \approx \exp(-1)$ and the fact that $\exp(-x) \leq 1/x$. To see why, it is enough to compute the probability that a single point does not fall into a ball of radius $2r$, then by independence we arrive at the joint probability above. That probability is 1 minus the probability that the point falls into the ball, which is itself:

$$\frac{\pi^{d/2}(2r)^d}{(d/2)!} = 2^d \underbrace{\frac{\pi^{d/2}r^d}{(d/2)!}}_{1/m},$$

hence the total probability $(1 - 2^d/m)^m$.

We have therefore shown that the probability that the distance of interest is r is extremely high and in the order of $1 - 1/2^d$, completing the proof. □

4.3 Randomized Trees

As we explained in Section 4.2, the search algorithm over a k-d Tree "index" consists of two operations: A single root-to-leaf traversal of the tree followed by backtracking to certify the candidate solution. As the analysis presented in the same section shows, it is the certification procedure that may need to visit virtually all data points. It is therefore not surprising that Liu et al. [2004] report that, in their experiments with low-dimensional vector collections (up to 30 dimensions), nearly 95% of the search time is spent in the latter phase.

That observation naturally leads to the following question: What if we eliminated the certification step altogether? In other words, when given a query q, the search algorithm simply finds the cell that contains q in $\mathcal{O}(\log m/m_\circ)$ time (where $m = |\mathcal{X}|$), then returns the solution from among the $m_\circ$ vectors in that cell—a strategy Liu et al. [2004] call *defeatist* search.

As Panigrahy [2008] shows for uniformly distributed vectors, however, the failure probability of the defeatist method is unacceptably high. That is primarily because, when a query is close to a decision boundary, the optimal solution may very well be on the other side. Figure 4.2(a) illustrates this phenomenon. As both the construction and search algorithms are *deterministic*, such a failure scenario is intrinsic to the algorithm and cannot be corrected once the tree has been constructed. Decision boundaries are hard and fast rules.

Would the situation be different if tree construction was a randomized algorithm? We could, for instance, let a random subset of the data points that are close to each decision boundary fall into both "left" and "right" sub-

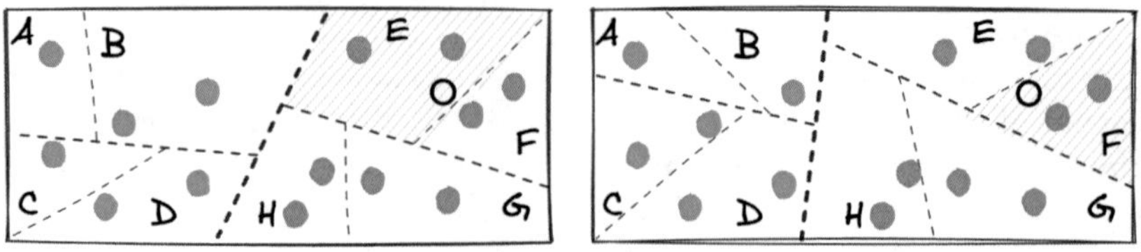

(a) (b)

Fig. 4.2: Randomized construction of k-d Trees for a fixed collection of vectors (filled circles). Decision boundaries take random directions and are planted at a randomly-chosen point near the median. Repeating this procedure results in multiple "index" structures of the vector collection. Performing a "defeatist" search repeatedly for a given query (the empty circles) then leads to a higher probability of success.

regions as we split an internal node. As another example, we could place the decision boundaries at randomly chosen points close to the median, and have them take a randomly chosen direction. We illustrate the latter in Figure 4.2.

Such randomized decisions mean that, every time we construct a k-d Tree, we would obtain a different index of the data. Furthermore, by building a forest of randomized k-d Trees and repeating the defeatist search algorithm, we may be able to lower the failure probability!

These, as we will learn in this section, are indeed successful ideas that have been extensively explored in the literature [Liu et al., 2004, Ram and Sinha, 2019, Dasgupta and Sinha, 2015].

4.3.1 Randomized Partition Trees

Recall that a decision boundary in a k-d Tree is an axis-aligned hyperplane that is placed at the median point of the projection of data points onto a coordinate axis. Consider the following adjustment to that procedure, due originally to Liu et al. [2004] and further refined by Dasgupta and Sinha [2015]. Every time a node whose region is $\mathcal{R}$ is to be split, we first draw a *random direction* u by sampling a vector from the d-dimensional unit sphere, and a scalar $\beta \in [1/4, 3/4]$ uniformly at random. We then project all data points that are in $\mathcal{R}$ onto u, and obtain the β-fractile of the projections, θ. u together with θ form the decision boundary. We then proceed as before to partition the data points in $\mathcal{R}$, by following the rule $\langle u, v\rangle \leq \theta$ for every data point $v \in \mathcal{R}$. A node turns into a leaf if it contains a maximum of $m_\circ$ vectors.

The procedure above gives us what Dasgupta and Sinha [2015] call a *Randomized Partition* (RP) Tree. You have already seen a visual demonstration of two RP Trees in Figure 4.2. Notice that, by requiring u to be a standard basis vector, fixing one u per each level of the tree, and letting $\beta = 0.5$, we reduce an RP Tree to the original k-d Tree.

What is the probability that a defeatist search over a single RP Tree fails to return the correct nearest neighbor? Dasgupta and Sinha [2015] proved that, that probability is related to the following potential function:

$$\Phi(q, \mathcal{X}) = \frac{1}{m} \sum_{i=1}^{m} \frac{\lVert q - x^{(\pi_1)} \rVert_2}{\lVert q - x^{(\pi_i)} \rVert_2}, \tag{4.3}$$

where $m = |\mathcal{X}|$, π_1 through π_m are indices that sort data points by increasing distance to the query point q, so that $x^{(\pi_1)}$ is the closest data point to q.

Notice that, a value of Φ that is close to 1 implies that nearly all data points are at the same distance from q. In that case, as we saw in Chapter 2, NN becomes unstable and approximate top-k retrieval becomes meaningless. When Φ is closer to 0, on the other hand, the optimal vector is well-separated from the rest of the collection.

Intuitively then, Φ is reflective of the difficulty or stability of the NN problem for a given query point. It makes sense then, that the probability of failure for q is related to this notion of difficulty of NN search: intuitively, when the nearest neighbor is far from other vectors, a defeatist search is more likely to yield the correct solution.

4.3.1.1 A Potential Function to Quantify the Difficulty of NN Search

Before we state the relationship between the failure probability and the potential function above more concretely, let us take a detour and understand where the expression for Φ comes from. All the arguments that we are about to make, including the lemmas and theorems, come directly from Dasgupta and Sinha [2015], though we repeat them here using our adopted notation for completeness. We also present an expanded proof of the formal results which follows the original proofs but elaborates on some of the steps.

Let us start with a simplified setup where $\mathcal{X}$ consists of just two vectors x and y. Suppose that for a query point q, $\lVert q - x \rVert_2 \leq \lVert q - y \rVert_2$. It turns out that, if we chose a random direction u and projected x and y onto it, then the probability that the projection of y onto u lands somewhere in between the projections of q and x onto u is a function of the potential function of Equation (4.3). The following lemma formalizes this relationship.

Lemma 4.2 *Suppose* $q, x, y \in \mathbb{R}^d$ *and* $\|q - x\|_2 \leq \|q - y\|_2$. *Let* $U \in \mathbb{S}^{d-1}$ *be a random direction and define* $\overset{\angle}{v} = \langle U, v \rangle$. *The probability that* $\overset{\angle}{y}$ *is between* $\overset{\angle}{q}$ *and* $\overset{\angle}{x}$ *is:*

$$\mathbb{P}\left[(\overset{\angle}{q} \leq \overset{\angle}{y} \leq \overset{\angle}{x}) \vee (\overset{\angle}{x} \leq \overset{\angle}{y} \leq \overset{\angle}{q})\right] =$$
$$\frac{1}{\pi} \arcsin\left(2\Phi(q, \{x, y\}) \sqrt{1 - \left(\frac{\langle q - x, y - x \rangle}{\|q - x\|_2 \|y - x\|_2} \right)^2} \right).$$

Proof. Assume, without loss of generality, that U is sampled from a d-dimensional standard Normal distribution: $U \sim \mathcal{N}(\mathbf{0}, I_d)$. That assumption is inconsequential because normalizing U by its L_2 norm gives a vector that lies on $\mathbb{S}^{d-1}$ as required. But because the norm of U does not affect the argument we need not explicitly perform the normalization.

Suppose further that we translate all vectors by q, and redefine $q \triangleq \mathbf{0}$, $x \triangleq x - q$, and $y \triangleq y - q$. We then rotate the vectors so that $x = \|x\|_2 e_1$, where e_1 is the first standard basis vector. Neither the translation nor the rotation affects pairwise distances, and as such, no generality is lost due to these transformations.

Given this arrangement of vectors, it will be convenient to write $U = (U_1, U_{\setminus 1})$ and $y = (y_1, y_{\setminus 1})$ so as to make explicit the first coordinate of each vector (denoted by subscript 1) and the remaining coordinates (denoted by subscript $\setminus 1$).

It is safe to assume that $y_{\setminus 1} \neq \mathbf{0}$. The reason is that, if that were not the case, the two vectors x and y have an intrinsic dimensionality of 1 and are thus on a line. In that case, no matter which direction U we choose, $\overset{\angle}{y}$ will not fall between $\overset{\angle}{x}$ and $\overset{\angle}{q} = \mathbf{0}$.

We can now write the probability of the event of interest as follows:

$$\mathbb{P}\Big[\underbrace{(\overset{\angle}{q} \leq \overset{\angle}{y} \leq \overset{\angle}{x}) \vee (\overset{\angle}{x} \leq \overset{\angle}{y} \leq \overset{\angle}{q})}_{E} \Big] =$$
$$\mathbb{P}\Big[\big(0 \leq \langle U, y \rangle \leq \|x\|_2 U_1\big) \vee \big(\|x\|_2 U_1 \leq \langle U, y \rangle \leq 0\big) \Big].$$

By expanding $\langle U, y \rangle = y_1 U_1 + \langle U_{\setminus 1}, y_{\setminus 1} \rangle$, it becomes clear that the expression above measures the probability that $\langle U_{\setminus 1}, y_{\setminus 1} \rangle$ falls in the interval $\big(-y_1|U_1|, (\|x\|_2 - y_1)|U_1| \big)$ when $U_1 \geq 0$ or $\big(-(\|x\|_2 - y_1)|U_1|, y_1|U_1| \big)$ otherwise. As such $\mathbb{P}[E]$ can be rewritten as follows:

$$\mathbb{P}[E] = \mathbb{P}\Big[-y_1|U_1| \leq \langle U_{\setminus 1}, y_{\setminus 1} \rangle \leq (\|x\|_2 - y_1)|U_1| \mid U_1 \geq 0 \Big] \mathbb{P}\big[U_1 \geq 0\big]$$
$$+ \mathbb{P}\Big[-(\|x\|_2 - y_1)|U_1| \leq \langle U_{\setminus 1}, y_{\setminus 1} \rangle \leq y_1|U_1| \mid U_1 < 0 \Big] \mathbb{P}\big[U_1 < 0\big].$$

First, note that U_1 is independent of $U_{\setminus 1}$ given that they are sampled from $\mathcal{N}(\mathbf{0}, I_d)$. Second, observe that $\langle U_{\setminus 1}, y_{\setminus 1} \rangle$ is distributed as $\mathcal{N}(\mathbf{0}, \|y_{\setminus 1}\|_2^2)$, which is symmetric, so that the two intervals have the same probability mass. These two observations simplify the expression above, so that $\mathbb{P}[E]$ becomes:

$$\begin{aligned}\mathbb{P}[E] &= \mathbb{P}\Big[-y_1|U_1| \leq \langle U_{\setminus 1}, y_{\setminus 1}\rangle \leq (\|x\|_2 - y_1)|U_1|\Big] \\ &= \mathbb{P}\Big[-y_1|Z| \leq \|y_{\setminus 1}\|_2 Z' \leq (\|x\|_2 - y_1)|Z|\Big] \\ &= \mathbb{P}\left[\frac{|Z'|}{|Z|} \in \Big(-\frac{y_1}{\|y_{\setminus 1}\|_2}, \frac{\|x\|_2 - y_1}{\|y_{\setminus 1}\|_2}\Big)\right],\end{aligned}$$

where Z and Z' are independent random variables drawn from $\mathcal{N}(0, 1)$.

Using the fact that the ratio of two independent Gaussian random variables follows a standard Cauchy distribution, we can calculate $\mathbb{P}[E]$ as follows:

$$\begin{aligned}\mathbb{P}[E] &= \int_{-y_1/\|y_{\setminus 1}\|_2}^{\big(\|x\|_2 - y_1\big)/\|y_{\setminus 1}\|_2} \frac{d\omega}{\pi(1+\omega^2)} \\ &= \frac{1}{\pi}\left[\arctan\Big(\frac{\|x\|_2 - y_1}{\|y_{\setminus 1}\|_2}\Big) - \arctan\Big(-\frac{y_1}{\|y_{\setminus 1}\|_2}\Big)\right] \\ &= \frac{1}{\pi}\arctan\left(\frac{\|x\|_2\|y_{\setminus 1}\|_2}{\|y\|_2^2 - y_1\|x\|_2}\right) \\ &= \frac{1}{\pi}\arcsin\left(\frac{\|x\|_2}{\|y\|_2}\sqrt{\frac{\|y\|_2^2 - y_1^2}{\|y\|_2^2 - 2y_1\|x\|_2 + \|x\|_2^2}}\right).\end{aligned}$$

In the third equality, we used the fact that $\arctan a + \arctan b = \arctan(a + b)/(1 - ab)$, and in the fourth equality we used the identity $\arctan a = \arcsin a/\sqrt{1+a^2}$. Substituting $y_1 = \langle y, x\rangle/\|x\|_2$ and noting that x and y have been shifted by q completes the proof. □

Corollary 4.1 *In the same configuration as in Lemma 4.2:*

$$\frac{2}{\pi}\Phi(q, \{x, y\})\sqrt{1 - \Big(\frac{\langle q - x, y - x\rangle}{\|q - x\|_2\|y - x\|_2}\Big)^2} \leq \mathbb{P}\Big[\big(\overset{\angle}{q} \leq \overset{\angle}{y} \leq \overset{\angle}{x}\big) \vee \big(\overset{\angle}{x} \leq \overset{\angle}{y} \leq \overset{\angle}{q}\big)\Big] \leq \Phi(q, \{x, y\})$$

Proof. Applying the inequality $\theta \geq \sin\theta \geq 2\theta/\pi$ for $0 \leq \theta \leq \pi/2$ to Lemma 4.2 implies the claim. □

Now that we have examined the case of $\mathcal{X} = \{x, y\}$, it is easy to extend the result to a configuration of m vectors.

Theorem 4.1 *Suppose $q \in \mathbb{R}^d$, $\mathcal{X} \subset \mathbb{R}^d$ is a set of m vectors, and $x^* \in \mathcal{X}$ the nearest neighbor of q. Let $U \in \mathbb{S}^{d-1}$ be a random direction, define $\overset{\angle}{v} = \langle U, v \rangle$, and let $\overset{\angle}{\mathcal{X}} = \{\overset{\angle}{x} \mid x \in \mathcal{X}\}$. Then:*

$$\mathbb{E}_U\big[\textit{fraction of } \overset{\angle}{\mathcal{X}} \textit{ that is between } \overset{\angle}{q} \textit{ and } \overset{\angle}{x^*}\big] \leq \frac{1}{2}\Phi(q, \mathcal{X}).$$

Proof. Let π_1 through π_m be indices that order the elements of $\mathcal{X}$ by increasing distance to q, so that $x^* = x^{(\pi_1)}$. Denote by Z_i the event that $\langle U, x^{(\pi_i)} \rangle$ falls between $\overset{\angle}{x^*}$ and $\overset{\angle}{q}$. By Corollary 4.1:

$$\mathbb{P}[Z_i] \leq \frac{1}{2} \frac{\|q - x^{(\pi_1)}\|_2}{\|q - x^{(\pi_i)}\|_2}.$$

We can now write the expectation of interest as follows:

$$\sum_{i=2}^{m} \frac{1}{m} \mathbb{P}[Z_i] \leq \frac{1}{2}\Phi(q, \mathcal{X}).$$

□

Corollary 4.2 *Under the assumptions of Theorem 4.1, for any $\alpha \in (0, 1)$ and any s-subset S of $\mathcal{X}$ that contains x^*:*

$$\mathbb{P}\left[\textit{at least } \alpha \textit{ fraction of } \overset{\angle}{S} \textit{ is between } \overset{\angle}{q} \textit{ and } \overset{\angle}{x^*}\right] \leq \frac{1}{2\alpha}\Phi_s(q, \mathcal{X}),$$

where:

$$\Phi_s(q, \mathcal{X}) = \frac{1}{s} \sum_{i=1}^{s} \frac{\|q - x^{(\pi_1)}\|_2}{\|q - x^{(\pi_i)}\|_2},$$

and π_1 through π_s are indices of the s vectors in $\mathcal{X}$ that are closest to q, ordered by increasing distance.

Proof. Apply Theorem 4.1 to the set S to obtain:

$$\mathbb{E}\big[\textit{fraction of } \overset{\angle}{S} \textit{ that is between } \overset{\angle}{q} \textit{ and } \overset{\angle}{x^*}\big] \leq \frac{1}{2}\Phi(q, S) \leq \frac{1}{2}\Phi_s(q, \mathcal{X}).$$

Using Markov's inequality (i.e., $\mathbb{P}[Z > \alpha] \leq \mathbb{E}[Z]/\alpha$) completes the proof. □

The above is where the potential function of Equation (4.3) first emerges in its complete form for an arbitrary collection of vectors and its subsets. As we see, Φ bounds the expected number of vectors whose projection onto a random direction U falls between a query point and its nearest neighbor.

The reason this expected value is important (which subsequently justifies the importance of Φ) has to do with the fact that decision boundaries are planted at some β-fractile point of the projections. As such, a bound on the number of points that fall between q and its nearest neighbor serves as a tool to bound the odds that the decision boundary may separate q from its nearest neighbor, which is the failure mode we wish to quantify.

4.3.1.2 Probability of Failure

We are now ready to use Theorem 4.1 and Corollary 4.2 to derive the failure probability of the defeatist search over an RP Tree. To that end, notice that, the path from the root to a leaf is a sequence of $\log_{1/\beta}(m/m_\circ)$ independent decisions that involve randomly projected data points. So if we were able to bound the failure probability of a single node, we can apply the union bound and obtain a bound on the failure probability of the tree. That is the intuition that leads to the following result.

Theorem 4.2 *The probability that an RP Tree built for collection $\mathcal{X}$ of m vectors fails to find the nearest neighbor of a query q is at most:*

$$\sum_{l=0}^{\ell} \Phi_{\beta^l m} \ln \frac{2e}{\Phi_{\beta^l m}},$$

with $\beta = 3/4$ and $\ell = \log_{1/\beta}(m/m_\circ)$, and where we use the shorthand Φ_s for $\Phi_s(q, \mathcal{X})$.

Proof. Consider an internal node of the RP Tree that contains q and s data points including x^*, the nearest neighbor of q. If the decision boundary at this node separates q from x^*, then the defeatist search will fail. We therefore seek to quantify the probability of that event.

Denote by F the fraction of the s vectors that, once projected onto the random direction U associated with the node, fall between q and x^*. Recall that, the split threshold associated with the node is drawn uniformly from an interval of mass $1/2$. As such, the probability that q is separated from x^* is at most $F/(1/2)$. By integrating over F, we obtain:

$$\begin{aligned}\mathbb{P}\left[q \textit{ is separated from } x^*\right] &\leq \int_0^1 \mathbb{P}\left[F = f\right]\frac{f}{1/2}df \\ &= 2\int_0^1 \mathbb{P}\left[F > f\right]df \\ &\leq 2\int_0^1 \min\left(1, \frac{\Phi_s}{2f}\right)df \\ &= 2\int_0^{\Phi_s/2} df + 2\int_{\Phi_s/2}^1 \frac{\Phi_s}{2f}df \\ &= \Phi_s \ln\frac{2e}{\Phi_s}.\end{aligned}$$

The first equality uses the definition of expectation for a positive random variable, while the second inequality uses Corollary 4.2. Applying the union bound to a path from root to leaf, and noting that the size of the collection that falls into each node drops geometrically per level by a factor of at least 3/4 completes the proof. □

We are thus able to express the failure probability as a function of Φ, a quantity that is defined for a particular q and a concrete collection of vectors. If we have a model of the data distribution, it may be possible to state more general bounds by bounding Φ itself. Dasgupta and Sinha [2015] demonstrate examples of this for two practical data distributions. Let us review one such example here.

4.3.1.3 Data Drawn from a Doubling Measure

Throughout our analysis of k-d Trees in Section 4.2, we considered the case where data points are uniformly distributed in $\mathbb{R}^d$. As we argued in Chapter 3, in many practical situations, however, even though vectors are represented in $\mathbb{R}^d$, they actually lie in some low-dimensional manifold with *intrinsic dimension* $d_\circ$ where $d_\circ \ll d$. This happens, for example, when data points are drawn from a *doubling measure* with low dimension as defined in Definition 3.1.

> Dasgupta and Sinha [2015] prove that, if a collection of m vectors is sampled from a doubling measure with dimension $d_\circ$, then Φ can be bounded from above roughly by $(1/m)^{1/d_\circ}$. The following theorem presents their claim.

Theorem 4.3 *Suppose a collection $\mathcal{X}$ of m vectors is drawn from μ, a continuous, doubling measure on $\mathbb{R}^d$ with dimension $d_\circ \geq 2$. For an arbitrary $\delta \in (0, 1/2)$, with probability at least $1 - 3\delta$, for all $2 \leq s \leq m$:*

$$\Phi_s(q, \mathcal{X}) \leq 6\left(\frac{2}{s}\ln\frac{1}{\delta}\right)^{1/d_\circ}.$$

Using the result above, Dasgupta and Sinha [2015] go on to prove that, under the same conditions, with probability at least $1-3\delta$, the failure probability of an RP Tree is bounded above by:

$$c_\circ(d_\circ + \ln m_\circ)\left(\frac{8\max(1, \ln 1/\delta)}{m_\circ}\right)^{1/d_\circ},$$

where $c_\circ$ is an absolute constant, and $m_\circ \geq c_\circ 3^{d_\circ}\max(1, \ln 1/\delta)$.

The results above tell us that, so long as the space has a small intrinsic dimension, we can make the probability of failing to find the optimal solution arbitrarily small.

4.3.2 Spill Trees

The Spill Tree [Liu et al., 2004] is another randomized variant of the k-d Tree. The algorithm to construct a Spill Tree comes with a hyperparameter $\alpha \in [0, 1/2]$ that is typically a small constant closer to 0. Given an α, the Spill Tree modifies the tree construction algorithm of the k-d Tree as follows. When splitting a node whose region is $\mathcal{R}$, we first project all vectors contained in $\mathcal{R}$ onto a random direction U, then find the median of the resulting distribution. However, instead of *partitioning* the vectors based on which side of the median they are on, the algorithm forms two overlapping sets. The "left" set contains all vectors in $\mathcal{R}$ whose projection onto U is smaller than the $(1/2+\alpha)$-fractile point of the distribution, while the "right" set consists of those that fall to the right of the $(1/2-\alpha)$-fractile point. As before, a node becomes a leaf when it has a maximum of $m_\circ$ vectors.

During search, the algorithm performs a defeatist search by routing the query point q based on a comparison of its projection onto the random direction associated with each node, and the *median* point. It is clear that with this strategy, if the nearest neighbor of q is close to the decision boundary of a node, we do not increase the likelihood of failure whether we route q to the left child or to the right one. Figure 4.3 shows an example of the defeatist search over a Spill Tree.

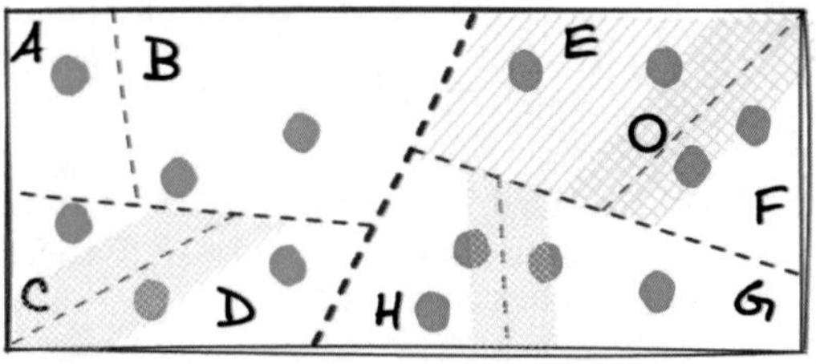

Fig. 4.3: Defeatist search over a Spill Tree. In a Spill Tree, vectors that are close to the decision boundary are, in effect, duplicated, with their copy "spilling" over to the other side of the boundary. This is depicted for a few example regions as the blue shaded area that straddles the decision boundary: vectors that fall into the shaded area belong to neighboring regions. For example, regions G and H share two vectors. As such, a defeatist search for the example query (the empty circle) looks through not just the region E but its extended region that overlaps with F.

4.3.2.1 Space Overhead

One obvious downside of the Spill Tree is that a single data point may end up in multiple leaf nodes, which increases the space complexity. We can quantify that by noting that the depth of the tree on a collection of m vectors is at most $\log_{1/(1/2+\alpha)}(m/m_\circ)$, so that the total number of vectors in all leaves is:

$$m_\circ 2^{\log_{1/(1/2+\alpha)}(m/m_\circ)} = m_\circ \big(\frac{m}{m_\circ}\big)^{\log_{1/(1/2+\alpha)} 2} = m_\circ \big(\frac{m}{m_\circ}\big)^{1/(1-\log(1+2\alpha))}.$$

As such, the space complexity of a Spill Tree is $\mathcal{O}(m^{1/(1-\log(1+2\alpha))})$.

4.3.2.2 Probability of Failure

The defeatist search over a Spill Tree fails to return the nearest neighbor x^* if the following event takes place at any of the nodes that contains q and x^*. That is the event where the projections of q and x^* are separated by the median *and* where the projection of x^* is separated from the median by at least α-fraction of the vectors. That event happens when the projections of q and x^* are separated by at least α-fraction of the vectors in some node along the path.

The probability of the event above can be bounded by Corollary 4.2. By applying the union bound to a root-to-leaf path, and noting that the size of the collection reduces at each level by a factor of at least $1/2 + \alpha$, we obtain the following result:

Theorem 4.4 *The probability that a Spill Tree built for collection $\mathcal{X}$ of m vectors fails to find the nearest neighbor of a query q is at most:*

$$\sum_{l=0}^{\ell} \frac{1}{2\alpha} \Phi_{\beta^l m}(q, \mathcal{X}),$$

with $\beta = 1/2 + \alpha$ and $\ell = \log_{1/\beta}\big(m/m_\circ\big)$.

4.4 Cover Trees

The branch-and-bound algorithms we have reviewed thus far divide a collection recursively into exactly two sub-collections, using a hyperplane as a decision boundary. Some also have a certification process that involves backtracking from a leaf node whose region contains a query to the root node. As we noted in Section 4.1, however, none of these choices is absolutely necessary. In fact, branching and bounding can be done entirely differently. We review in this section a popular example that deviates from that pattern, a data structure known as the Cover Tree [Beygelzimer et al., 2006].

It is more intuitive to describe the Cover Tree, as well as the construction and search algorithms over it, in the abstract first. This is what Beygelzimer et al. [2006] call the *implicit* representation. Let us first describe its structure, then review its properties and explain the relevant algorithms, and only then discuss how the abstract tree can be implemented concretely.

4.4.1 The Abstract Cover Tree and its Properties

The abstract Cover Tree is a tree structure with infinite depth that is defined for a proper metric $\delta(\cdot, \cdot)$. Each level of the tree is numbered by an integer that starts from ∞ at the level of the root node and decrements, to $-\infty$, at each subsequent level. Each node represents a single data point. If we denote the collection of nodes on level ℓ by C_ℓ, then C_ℓ is a *set*, in the sense that the data points represented by those nodes are distinct. But $C_\ell \subset C_{\ell-1}$, so that once a node appears in level ℓ, it necessarily appears in levels $(\ell - 1)$ onward. That implies that, in the abstract Cover Tree, C_∞ contains a single data point, and $C_{-\infty} = \mathcal{X}$ is the entire collection.

This structure, which is illustrated in Figure 4.4 for an example collection of vectors, obeys three invariants. That is, all algorithms that construct the tree or manipulate it in any way must guarantee that the three properties are not violated. These invariants are:

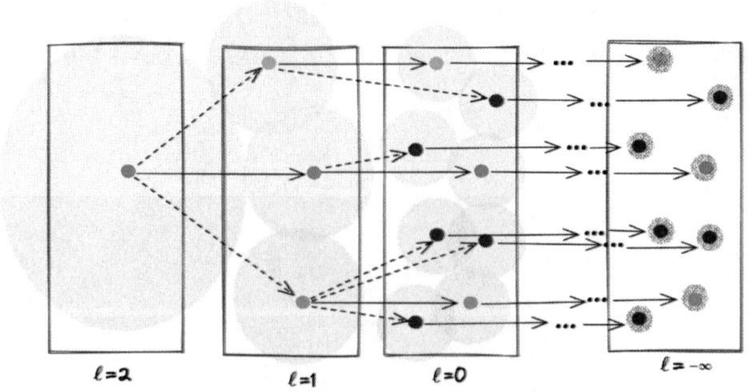

Fig. 4.4: Illustration of the abstract Cover Tree for a collection of 8 vectors. Nodes on level ℓ of the tree are separated by at least 2^ℓ by the separation invariant. Nodes on level ℓ cover nodes on level $(\ell - 1)$ with a ball of radius at most 2^ℓ by the covering invariant. Once a node appears in the tree, it will appear on all subsequent levels as its own child (solid arrows), by the nesting invariant.

- **Nesting**: As we noted, $C_\ell \subset C_{\ell-1}$.
- **Covering**: For every node $u \in C_{\ell-1}$ there is a node $v \in C_\ell$ such that $\delta(u, v) < 2^\ell$. In other words, every node in the next level $(\ell - 1)$ of the tree is "covered" by an open ball of radius 2^ℓ around a node in the current level, ℓ.
- **Separation**: All nodes on the same level ℓ are separated by a distance of at least 2^ℓ. Formally, if $u, v \in C_\ell$, then $\delta(u, v) > 2^\ell$.

4.4.2 The Search Algorithm

We have seen what a Cover Tree looks like and what properties it is guaranteed to maintain. Given this structure, how do we find the nearest neighbor of a query point? That turns out to be a fairly simple algorithm as shown in Algorithm 1.

Algorithm 1 always maintains a current set of candidates in Q_ℓ as it visits level ℓ of the tree. In each iteration of the loop on Line 2, it creates a temporary set—denoted by Q—by collecting the children of all nodes in Q_ℓ. It then prunes the nodes in Q based on the condition on Line 4. Eventually, the algorithm returns the exact nearest neighbor of query q by performing exhaustive search over the nodes in $Q_{-\infty}$.

Let us understand why the algorithm is correct. In a way, it is enough to argue that the pruning condition on Line 4 never discards an ancestor of the nearest neighbor. If that were the case, we are done proving the correctness

Algorithm 1: Nearest Neighbor search over a Cover Tree.

Input: Cover Tree with metric $\delta(\cdot,\cdot)$; query point q.
Result: Exact NN of q.
1: $Q_\infty \leftarrow C_\infty$; ▷ C_ℓ `is the set of nodes on level` ℓ
2: **for** ℓ from ∞ to $-\infty$ **do**
3: $Q \leftarrow \{\text{CHILDREN}(v) \mid v \in Q_\ell\}$; ▷ `CHILDREN(·) returns the children of its argument.`
4: $Q_{\ell-1} \leftarrow \{u \mid \delta(q,u) \leq \delta(q,Q) + 2^\ell\}$; ▷ $\delta(u,S) \triangleq \min_{v\in S}\delta(u,v)$.
5: **end for**
6: **return** $\arg\min_{u\in Q_{-\infty}} \delta(q,u)$

of the algorithm: $Q_{-\infty}$ is guaranteed to have the nearest neighbor, at which point we will find it on Line 6.

The fact that Algorithm 1 never prunes the ancestor of the solution is easy to establish. To see how, consider the distance between $u \in C_{\ell-1}$ and any of its descendants, v. The distance between the two vectors is bounded as follows: $\delta(u,v) \leq \sum_{l=\ell-1}^{-\infty} 2^l = 2^\ell$. Furthermore, because δ is proper, by triangle inequality, we know that: $\delta(q,u^*) \leq \delta(q,Q) + \delta(Q,u^*)$, where u^* is the solution and a descendant of $u \in C_{\ell-1}$. As such, any candidate whose distance is greater than $\delta(q,Q)+\delta(Q,u^*) \leq \delta(q,Q)+2^\ell$ can be safely pruned.

The search algorithm has an ϵ-approximate variant too. To obtain a solution that is at most $(1+\epsilon)\delta(q,u^*)$ away from q, assuming u^* is the optimal solution, we need only to change the termination condition on Line 2, by exiting the loop as soon as $\delta(q,Q_\ell) \geq 2^{\ell+1}(1+1/\epsilon)$. Let us explain why the resulting algorithm is correct.

Suppose that the algorithm terminates early when it reaches level ℓ. That means that $2^{\ell+1}(1+1/\epsilon) \leq \delta(q,Q_\ell)$. We have already seen that $\delta(q,Q_\ell) \leq 2^{\ell+1}$, and by triangle inequality, that $\delta(q,Q_\ell) \leq \delta(q,u^*) + 2^{\ell+1}$. So we have bounded $\delta(q,Q_\ell)$ from below and above, resulting in the following inequality:

$$2^{\ell+1}\left(1+\frac{1}{\epsilon}\right) \leq \delta(q,u^*) + 2^{\ell+1} \implies 2^{\ell+1} \leq \epsilon\delta(q,u^*).$$

Putting all that together, we have shown that $\delta(q,Q_\ell) \leq (1+\epsilon)\delta(q,u^*)$, so that Line 6 returns an ϵ-approximate solution.

4.4.3 The Construction Algorithm

Inserting a single vector into the Cover Tree "index" is a procedure that is similar to the search algorithm but is better conceptualized recursively, as shown in Algorithm 2.

Algorithm 2: Insertion of a vector into a Cover Tree.

Input: Cover Tree $\mathcal{T}$ with metric $\delta(\cdot, \cdot)$; New Vector p; Level ℓ; Candidate set Q_ℓ.
Result: Cover Tree containing p.

1: $Q \leftarrow \{\textsc{Children}(u) \mid u \in Q_\ell\}$
2: **if** $\delta(p, Q) > 2^\ell$ **then**
3:     **return** $\bowtie$
4: **else**
5:     $Q_{\ell-1} \leftarrow \{u \in Q \mid \delta(p, u) \leq 2^\ell\}$
6:     **if Insert**$(\mathcal{T}, p, Q_{\ell-1}, \ell - 1) = \bowtie \wedge \delta(p, Q_\ell) \leq 2^\ell$ **then**
7:         Choose $u \in Q_\ell$ such that $\delta(p, u) \leq 2^\ell$
8:         Add p to $\textsc{Children}(u)$
9:         **return** $\blacklozenge$
10:     **else**
11:         **return** $\bowtie$
12:     **end if**
13: **end if**

It is important to note that the procedure in Algorithm 2 assumes that the point p is not present in the tree. That is a harmless assumption as the existence of p can be checked by a simple invocation of Algorithm 1. We can therefore safely assume that $\delta(p, Q)$ for any Q formed on Line 1 is strictly positive. That assumption guarantees that the algorithm eventually terminates. That is because $\delta(p, Q) > 0$ so that ultimately we will invoke the algorithm with a value ℓ such that $\delta(p, Q) > 2^\ell$, at which point Line 2 terminates the recursion.

We can also see why Line 6 is bound to evaluate to True at some point during the execution of the algorithm. That is because there must exist a level ℓ such that $2^{\ell-1} < \delta(p, Q) \leq 2^\ell$. That implies that the point p will ultimately be inserted into the tree.

What about the three invariants of the Cover Tree? We must now show that the resulting tree maintains those properties: nesting, covering, and separation. The covering invariant is immediately guaranteed as a result of Line 6. The nesting invariant too is trivially maintained because we can insert p as its own child for all subsequent levels.

What remains is to show that the insertion algorithm maintains the separation property. To that end, suppose p has been inserted into $C_{\ell-1}$ and consider its sibling $u \in C_{\ell-1}$. If $u \in Q$, then it is clear that $\delta(p, u) > 2^{\ell-1}$ because Line 6 must have evaluated to True. On the other hand, if $u \notin Q$, that means that there was some $\ell' > \ell$ where some ancestor of u, $u' \in C_{\ell'-1}$, was pruned on Line 5, so that $\delta(p, u') > 2^{\ell'}$. Using the covering invariant, we can deduce that:

$$\begin{aligned}\delta(p, u) &\geq \delta(p, u') - \sum_{l=\ell'-1}^{\ell} 2^l \\ &= \delta(p, u') - (2^{\ell'} - 2^{\ell}) \\ &> 2^{\ell'} - (2^{\ell'} - 2^{\ell}) = 2^{\ell}.\end{aligned}$$

That concludes the proof that $\delta(p, C_{\ell-1}) > 2^{\ell-1}$, showing that Algorithm 2 maintains the separation invariant.

4.4.4 The Concrete Cover Tree

The abstract tree we described earlier has infinite depth. While that representation is convenient for explaining the data structure and algorithms that operate on it, it is not practical. But it is easy to derive a concrete instance of the data structure, without changing the algorithmic details, to obtain what Beygelzimer et al. [2006] call the *explicit* representation.

One straightforward way of turning the abstract Cover Tree into a concrete one is by turning a node into a (terminal) leaf if it is its only child—recall that, a node in the abstract Cover Tree is its own child, indefinitely. For example, in Figure 4.4, all nodes on level 0 would become leaves and the Cover Tree would end at that depth. We leave it as an exercise to show that the concrete representation of the tree does not affect the correctness of Algorithms 1 and 2.

The concrete form is not only important for making the data structure practical, it is also necessary for analysis. For example, as Beygelzimer et al. [2006] prove that the space complexity of the concrete Cover Tree is $\mathcal{O}(m)$ with $m = |\mathcal{X}|$, whereas the abstract form is infinitely large. The time complexity of the insertion and search algorithms also use the concrete form, but they further require assumptions on the data distribution. Beygelzimer et al. [2006] present their analysis for vectors that are drawn from a doubling measure, as we have defined in Definition 3.1. However, their claims have been disputed [Curtin, 2016] by counter-examples [Elkin and Kurlin, 2022], and corrected in a recent work [Elkin and Kurlin, 2023].

4.5 Closing Remarks

This chapter has only covered algorithms that convey the foundations of a branch-and-bound approach to NN search. Indeed, we left out a number of alternative constructions that are worth mentioning as we close this chapter.

4.5.1 Alternative Constructions and Extensions

The standard k-d Tree itself, as an example, can be instantiated by using a different splitting procedure, such as splitting on the axis along which the data exhibits the greatest spread. PCA Trees [Sproull, 1991], PAC Trees [McNames, 2001], and Max-Margin Trees [Ram et al., 2012] offer other ways of choosing the axis or direction along which the algorithm partitions the data. Vantage-point Trees [Yianilos, 1993], as another example, follow the same iterative procedure as k-d Trees, but partition the space using hyperspheres rather than hyperplanes.

There are also various other randomized constructions of tree index structures for NN search. Panigrahy [2008], for instance, construct a standard k-d Tree over the original data points but, during search, perturb the query point. Repeating the perturb-then-search scheme reduces the failure probability of a defeatist search over the k-d Tree.

Sinha [2014] proposes a different variant of the RP Tree where, instead of a random projection, they choose the principal direction corresponding to the largest eigenvalue of the covariance of the vectors that fall into a node. This is equivalent to the PAC Tree [McNames, 2001] with the exception that the splitting threshold (i.e., the β-fractile point) is chosen randomly, rather than setting it to the median point. Sinha [2014] shows that, with the modified algorithm, a smaller ensemble of trees is necessary to reach high retrieval accuracy, as compared with the original RP Tree construction.

Sinha and Keivani [2017] improve the space complexity of RP Trees by replacing the d-dimensional *dense* random direction with a *sparse* random projection using Fast Johnson-Lindenstrauss Transform [Ailon and Chazelle, 2009]. The result is that, every internal node of the tree has to store a sparse vector whose number of non-zero coordinates is far less than d. This space-efficient variant of the RP Tree offers virtually the same theoretical guarantees as the original RP Tree structure.

Ram and Sinha [2019] improve the running time of the NN search over an RP Tree (which is $\mathcal{O}(d \log m)$ for $m = |\mathcal{X}|$) by first randomly rotating the vectors in a pre-processing step, then applying the standard k-d Tree to the rotated vectors. They show that, such a construction leads to a search time complexity of $\mathcal{O}(d \log d + \log m)$ and offers the same guarantees on the failure probability as the RP Tree.

Cover Trees too have been the center of much research. As we have already mentioned, many subsequent works [Elkin and Kurlin, 2022, 2023, Curtin, 2016] investigated the theoretical results presented in the original paper [Beygelzimer et al., 2006] and corrected or improved the time complexity bounds on the insertion and search algorithms. Izbicki and Shelton [2015] simplified the structure of the concrete Cover Tree to make its implementation more efficient and cache-aware. Gu et al. [2022] proposed parallel insertion and deletion algorithms for the Cover Tree to scale the algorithm to real-world vector collections. We should also note that the Cover Tree itself is an extension (or, rather, a simplification) of Navigating Nets [Krauthgamer and Lee, 2004], which itself has garnered much research.

It is also possible to extend the framework to MIPS. That may be surprising. After all, the machinery of the branch-and-bound framework rests on the assumption that the distance function has all the nice properties we expect from a metric space. In particular, we take for granted that the distance is non-negative and that distances obey the triangle inequality. As we know, however, none of these properties holds when the distance function is inner product.

As Bachrach et al. [2014] show, however, it is possible to apply a rank-preserving transformation to vectors such that solving MIPS over the original space is equivalent to solving NN over the transformed space. Ram and Gray [2012] take a different approach and derive bounds on the inner product between an arbitrary query point and vectors that are contained in a ball associated with an internal node of the tree index. This bound allows the certification process to proceed as usual. Nonetheless, these methods face the same challenges as k-d Trees and their variants.

4.5.2 Future Directions

The literature on branch-and-bound algorithms for top-k retrieval is rather mature and stable at the time of this writing. While publications on this fascinating class of algorithms continue to date, most recent works either improve the theoretical analysis of existing algorithms (e.g., [Elkin and Kurlin, 2023]), improve their implementation (e.g., [Ram and Sinha, 2019]), or adapt their implementation to other computing paradigms such as distributed systems (e.g., [Gu et al., 2022]).

Indeed, such research is essential. Tree indices are—as the reader will undoubtedly learn after reading this monograph—among the few retrieval algorithms that rest on a sound theoretical foundation. Crucially, their implementations too reflect those theoretical principles: There is little to no gap between theoretical tree indices and their concrete forms. Improving their theoretical guarantees and modernizing their implementation, therefore, makes a great deal of sense, especially so because works like [Ram and Sinha, 2019] show how competitive tree indices can be in practice.

An example area that has received little attention concerns the data structure that materializes a tree index. In most works, trees appear in their naïve form and are processed trivially. That is, a tree is simply a collection of if-else blocks, and is evaluated from root to leaf, one node at a time. The vectors in the leaf of a tree, too, are simply searched exhaustively. Importantly, the knowledge that one tree is often insufficient and that a forest of trees is often necessary to reach an acceptable retrieval accuracy, is not taken advantage of. This insight was key in improving forest traversal in the learning-to-rank literature [Lucchese et al., 2015, Ye et al., 2018], in particular when a batch of queries is to be processed simultaneously. It remains to be seen if a more efficient tree traversal algorithm can unlock the power of tree indices.

Perhaps more importantly, the algorithms we studied in this chapter give us an arsenal of theoretical tools that may be of independent interest. The concepts such as partitioning, spillage, and ϵ-nets that are so critical in the development of many of the algorithms we saw earlier, are useful not only in the context of trees, but also in other classes of retrieval algorithms. We will say more on that in future chapters.

References

N. Ailon and B. Chazelle. The fast johnson–lindenstrauss transform and approximate nearest neighbors. *SIAM Journal on Computing*, 39(1):302–322, 2009.

Y. Bachrach, Y. Finkelstein, R. Gilad-Bachrach, L. Katzir, N. Koenigstein, N. Nice, and U. Paquet. Speeding up the xbox recommender system using a euclidean transformation for inner-product spaces. In *Proceedings of the 8th ACM Conference on Recommender Systems*, page 257–264, 2014.

J. L. Bentley. Multidimensional binary search trees used for associative searching. *Communications of the ACM*, 18(9):509–517, 9 1975.

A. Beygelzimer, S. Kakade, and J. Langford. Cover trees for nearest neighbor. In *Proceedings of the 23rd International Conference on Machine Learning*, page 97–104, 2006.

P. Ciaccia, M. Patella, and P. Zezula. M-tree: An efficient access method for similarity search in metric spaces. In *Proceedings of the 23rd International Conference on Very Large Data Bases*, page 426–435, 1997.

K. L. Clarkson. Nearest neighbor queries in metric spaces. In *Proceedings of the Twenty-Ninth Annual ACM Symposium on Theory of Computing*, pages 609–617, 1997.

R. R. Curtin. *Improving dual-tree algorithms.* PhD thesis, Georgia Institute of Technology, Atlanta, GA, USA, 2016.

S. Dasgupta and K. Sinha. Randomized partition trees for nearest neighbor search. *Algorithmica*, 72(1):237–263, 5 2015.

Y. Elkin and V. Kurlin. Counterexamples expose gaps in the proof of time complexity for cover trees introduced in 2006. In *2022 Topological Data Analysis and Visualization*, pages 9–17, Los Alamitos, CA, 10 2022.

Y. Elkin and V. Kurlin. A new near-linear time algorithm for k-nearest neighbor search using a compressed cover tree. In *Proceedings of the 40th International Conference on Machine Learning*, 2023.

J. H. Friedman, J. L. Bentley, and R. A. Finkel. An algorithm for finding best matches in logarithmic expected time. *ACM Transactions on Mathematical Software*, 3(3):209–226, 9 1977.

Y. Gu, Z. Napier, Y. Sun, and L. Wang. Parallel cover trees and their applications. In *Proceedings of the 34th ACM Symposium on Parallelism in Algorithms and Architectures*, pages 259–272, 2022.

M. Izbicki and C. Shelton. Faster cover trees. In *Proceedings of the 32nd International Conference on Machine Learning*, volume 37 of *Proceedings of Machine Learning Research*, pages 1162–1170, Lille, France, 07–09 Jul 2015.

D. R. Karger and M. Ruhl. Finding nearest neighbors in growth-restricted metrics. In *Proceedings of the 34th Annual ACM Symposium on Theory of Computing*, pages 741–750, 2002.

R. Krauthgamer and J. R. Lee. Navigating nets: Simple algorithms for proximity search. In *Proceedings of the 15th Annual ACM-SIAM Symposium on Discrete Algorithms*, pages 798–807, 2004.

T. Liu, A. W. Moore, A. Gray, and K. Yang. An investigation of practical approximate nearest neighbor algorithms. In *Proceedings of the 17th International Conference on Neural Information Processing Systems*, pages 825–832, 2004.

C. Lucchese, F. M. Nardini, S. Orlando, R. Perego, N. Tonellotto, and R. Venturini. Quickscorer: A fast algorithm to rank documents with additive ensembles of regression trees. In *Proceedings of the 38th International ACM SIGIR Conference on Research and Development in Information Retrieval*, pages 73–82, 2015.

J. McNames. A fast nearest-neighbor algorithm based on a principal axis search tree. *IEEE Transactions on Pattern Analysis and Machine Intelligence*, 23(9):964–976, 2001.

R. Panigrahy. An improved algorithm finding nearest neighbor using kd-trees. In *LATIN 2008: Theoretical Informatics*, pages 387–398, 2008.

P. Ram and A. G. Gray. Maximum inner-product search using cone trees. In *Proceedings of the 18th ACM SIGKDD International Conference on Knowledge Discovery and Data Mining*, page 931–939, 2012.

P. Ram and K. Sinha. Revisiting kd-tree for nearest neighbor search. In *Proceedings of the 25th ACM SIGKDD International Conference on Knowledge Discovery and Data Mining*, pages 1378–1388, 2019.

P. Ram, D. Lee, and A. G. Gray. Nearest-neighbor search on a time budget via max-margin trees. In *Proceedings of the 2012 SIAM International Conference on Data Mining*, pages 1011–1022, 2012.

K. Sinha. Lsh vs randomized partition trees: Which one to use for nearest neighbor search? In *Proceedings of the 13th International Conference on Machine Learning and Applications*, pages 41–46, 2014.

K. Sinha and O. Keivani. Sparse Randomized Partition Trees for Nearest Neighbor Search. In *Proceedings of the 20th International Conference on*

Artificial Intelligence and Statistics, volume 54 of *Proceedings of Machine Learning Research*, pages 681–689, 20–22 Apr 2017.

R. F. Sproull. Refinements to nearest-neighbor searching ink-dimensional trees. *Algorithmica*, 6(1):579–589, 6 1991.

T. Ye, H. Zhou, W. Y. Zou, B. Gao, and R. Zhang. Rapidscorer: Fast tree ensemble evaluation by maximizing compactness in data level parallelization. In *Proceedings of the 24th ACM SIGKDD International Conference on Knowledge Discovery and Data Mining*, pages 941–950, 2018.

P. N. Yianilos. Data structures and algorithms for nearest neighbor search in general metric spaces. In *Proceedings of the 4th Annual ACM-SIAM Symposium on Discrete Algorithms*, pages 311–321, 1993.

Chapter 5
Locality Sensitive Hashing

Abstract In the preceding chapter, we delved into algorithms that inferred the geometrical shape of a collection of vectors and condensed it into a navigable structure. In many cases, the algorithms were designed for exact top-k retrieval, but could be modified to provide guarantees on approximate search. This section, instead, explores an entirely different idea that is probabilistic in nature and, as such, is designed specifically for approximate top-k retrieval from the ground up.

5.1 Intuition

Let us consider the intuition behind what is known as *Locality Sensitive Hashing* (LSH) [Indyk and Motwani, 1998] first. Define b separate "buckets." Now, suppose there exists a mapping $h(\cdot)$ from vectors in $\mathbb{R}^d$ to these buckets, such that every vector is placed into a single bucket: $h : \mathbb{R}^d \rightarrow [b]$. Crucially, assume that vectors that are closer to each other according to the distance function $\delta(\cdot, \cdot)$, are more likely to be placed into the same bucket. In other words, the probability that two vectors collide increases as δ decreases.

Considering the setup above, indexing is simply a matter of applying h to all vectors in the collection $\mathcal{X}$ and making note of the resulting placements. Retrieval for a query q is also straightforward: Perform exact search over the data points that are in the bucket $h(q)$. The reason this procedure works with high probability is because it is more likely for the mapping h to place q in a bucket that contains its nearest neighbors, so that an exact search over the $h(q)$ bucket yields the correct top-k vectors with high likelihood. This is visualized in Figure 5.1(a).

It is easy to extend this setup to "multi-dimensional" buckets in the following sense. If h_i's are independent functions that have the desired property above (i.e., increased chance of collision with smaller δ), we may define a bucket in $[b]^\ell$ as the vector mapping $g(\cdot) = [h_1(\cdot), h_2(\cdot), \ldots, h_\ell(\cdot)]$. Fig-

S. Bruch, *Foundations of Vector Retrieval*, https://doi.org/10.1007/978-3-031-55182-6_5

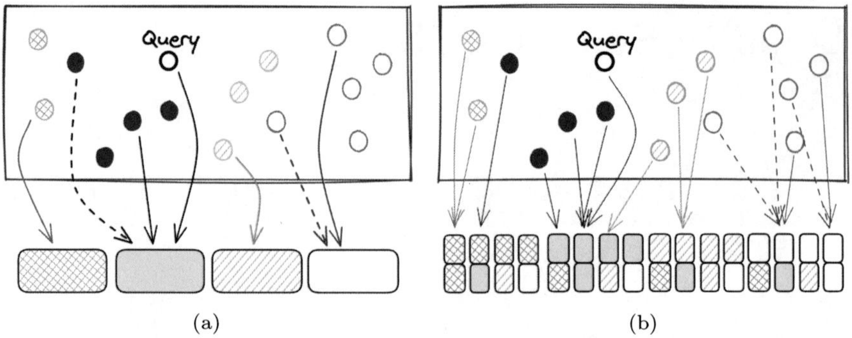

Fig. 5.1: Illustration of Locality Sensitive Hashing. In (a), a function $h : \mathbb{R}^2 \rightarrow \{1, 2, 3, 4\}$ maps vectors to four buckets. Ideally, when two points are closer to each other, they are more likely to be placed in the same bucket. But, as the dashed arrows show, some vectors end up in less-than-ideal buckets. When retrieving the top-k vectors for a query q, we search through the data vectors that are in the bucket $h(q)$. Figure (b) depicts an extension of the framework where each bucket is the vector $[h_1(\cdot), h_2(\cdot)]$ obtained from two independent mappings h_1 and h_2.

ure 5.1(b) illustrates this extension for $\ell = 2$. The indexing and search procedures work in much the same way. But now, there are presumably fewer data points in each bucket, and spurious collisions (i.e., vectors that were mapped to the same bucket but that are far from each other according to δ) are less likely to occur. In this way, we are likely to reduce the overall search time and increase the accuracy of the algorithm.

Extending the framework even further, we can repeat the process above L times by constructing independent mappings $g_1(\cdot)$ through $g_L(\cdot)$ from individual mappings $h_{ij}(\cdot)$ ($1 \leq i \leq L$ and $1 \leq j \leq \ell$), all of which possessing the property of interest. Because the mappings are independent, repeating the procedure many times increases the probability of obtaining a high retrieval accuracy.

That is the essence of the LSH approach to top-k retrieval. Its key ingredient is the family $\mathcal{H}$ of functions h_{ij}'s that have the stated property *for a given distance function*, δ. This is the detail that is studied in the remainder of this section. But before we proceed to define $\mathcal{H}$ for different distance functions, we will first give a more rigorous description of the algorithm.

5.2 Top-k Retrieval with LSH

Earlier, we described informally the class of mappings that are at the core of LSH, as hash functions that preserve the distance between points. That is, the likelihood that such a hash function places two points in the same bucket is a function of their distance. Let us formalize that notion first in the following definition, due to Indyk and Motwani [1998].

Definition 5.1 ($(r, (1+\epsilon)r, p_1, p_2)$-Sensitive Family) A family of hash functions $\mathcal{H} = \{h : \mathbb{R}^d \rightarrow [b]\}$ is called $(r, (1+\epsilon)r, p_1, p_2)$-sensitive for a distance function $\delta(\cdot, \cdot)$, where $\epsilon > 0$ and $0 < p_1, p_2 < 1$, if for any two points $u, v \in \mathbb{R}^d$:

- $\delta(u, v) \leq r \implies \mathbb{P}_{\mathcal{H}}\big[h(u) = h(v)\big] \geq p_1$; and,
- $\delta(u, v) > (1+\epsilon)r \implies \mathbb{P}_{\mathcal{H}}\big[h(u) = h(v)\big] \leq p_2$.

It is clear that such a family is useful only when $p_1 > p_2$. We will see examples of $\mathcal{H}$ for different distance functions later in this section. For the time being, however, suppose such a family of functions exists for any δ of interest.

The indexing algorithm remains as described before. Fix parameters ℓ and L to be determined later in this section. Then define the vector function $g(\cdot) = \big[h_1(\cdot), h_2(\cdot), \ldots, h_\ell(\cdot)\big]$ where $h_i \in \mathcal{H}$. Now, construct L such functions g_1 through g_L, and process the data points in collection $\mathcal{X}$ by evaluating g_i's and placing them in the corresponding multi-dimensional bucket.

In the end, we have effectively built L tables, each mapping buckets to a list of data points that fall into them. Note that, each of the L tables holds a copy of the collection, but where each table organizes the data points differently.

5.2.1 The Point Location in Equal Balls Problem

Our intuitive description of retrieval using LSH ignored a minor technicality that we must elaborate in this section. In particular, as is clear from Definition 5.1, a family $\mathcal{H}$ has a dependency on the distance r. That means any instance of the family provides guarantees only with respect to a specific r. Consequently, any index obtained from a family $\mathcal{H}$, too, is only useful in the context of a fixed r.

It appears, then, that the LSH index is not in and of itself sufficient for solving the ϵ-approximate retrieval problem of Definition 1.2 directly. But, it is enough for solving an easier *decision problem* that is known as Point Location in Equal Balls (PLEB), defined as follows:

Definition 5.2 ($(r,(1+\epsilon)r)$-Point Location in Equal Balls) For a query point q and a collection $\mathcal{X}$, if there is a point $u \in \mathcal{X}$ such that $\delta(q,u) \leq r$, return YES and any point v such that $\delta(q,v) < (1+\epsilon)r$. Return NO if there are no such points.

The algorithm to solve the $(r,(1+\epsilon)r)$-PLEB problem for a query point q is fairly straightforward. It involves evaluating g_i's on q and exhaustively searching the corresponding buckets in order. We may terminate early after visiting at most $4L$ data points. For every examined data point u, the algorithm returns YES if $\delta(q,u) \leq (1+\epsilon)r$, and NO otherwise.

5.2.1.1 Proof of Correctness

Suppose there exits a point $u^* \in \mathcal{X}$ such that $\delta(q,u^*) \leq r$. The algorithm above is correct, in the sense that it returns a point u with $\delta(q,u) \leq (1+\epsilon)r$, if we choose ℓ and L such that the following two properties hold with constant probability:

- $\exists\, i \in [L]$ *s.t.* $g_i(u^*) = g_i(q)$; and,
- $\sum_{j=1}^{L} \Big| \Big(\mathcal{X} \setminus B(q,(1+\epsilon)r)\Big) \cap g_j^{-1}(g_j(q)) \Big| \leq 4L$, where $g_j^{-1}(g_j(q))$ is the set of vectors in bucket $g_j(q)$.

The first property ensures that, as we traverse the L buckets associated with the query point, we are likely to visit either the optimal point u^*, or some other point whose distance to q is at most $(1+\epsilon)r$. The second property guarantees that with constant probability, there are no more than $4L$ points in the candidate buckets that are $(1+\epsilon)r$ away from q. As such, we are likely to find a solution before visiting $4L$ points.

We must therefore prove that for some ℓ and L the above properties hold. The following claim shows one such configuration.

Theorem 5.1 *Let $\rho = \ln p_1 / \ln p_2$ and $m = |\mathcal{X}|$. Set $L = m^\rho$ and $\ell = \log_{1/p_2} m$. The properties above hold with constant probability for a $(r,(1+\epsilon)r,p_1,p_2)$-sensitive LSH family.*

Proof. Consider the first property. We have, from Definition 5.1, that, for any $h_i \in \mathcal{H}$:

$$\mathbb{P}\Big[h_i(u^*) = h_i(q)\Big] \geq p_1.$$

That holds simply because $u^* \in B(q,r)$. That implies:

$$\mathbb{P}\Big[g_i(u^*) = g_i(q)\Big] \geq p_1^\ell.$$

As such:

$$\mathbb{P}\Big[\exists\, i \in [L] \text{ s.t. } g_i(u^*) = g_i(q)\Big] \geq 1 - (1-p_1^\ell)^L.$$

Substituting ℓ and L with the expressions given in the theorem gives:

$$\mathbb{P}\Big[\exists\, i \in [L] \ s.t.\ g_i(u^*) = g_i(q)\Big] \geq 1 - (1 - \frac{1}{m^\rho})^{m^\rho} \approx 1 - \frac{1}{e},$$

proving that the property of interest holds with constant probability.

Next, consider the second property. For any point v such that $\delta(q, v) > (1 + \epsilon)r$, Definition 5.1 tells us that:

$$\begin{aligned}
\mathbb{P}\Big[h_i(v) = h_i(q)\Big] \leq p_2 &\implies \mathbb{P}\Big[g_i(v) = g_i(q)\Big] \leq p_2^\ell \\
&\implies \mathbb{P}\Big[g_i(v) = g_i(q)\Big] \leq \frac{1}{m} \\
&\implies \mathbb{E}\Big[\Big|v \ s.t.\ g_i(v) = g_i(q) \wedge \delta(q, v) > (1 + \epsilon)r\Big| \mid g_i\Big] \leq 1 \\
&\implies \mathbb{E}\Big[\Big|v \ s.t.\ g_i(v) = g_i(q) \wedge \delta(q, v) > (1 + \epsilon)r\Big|\Big] \leq L,
\end{aligned}$$

where the last expression follows by the linearity of expectation when applied to all L buckets. By Markov's inequality, the probability that there are more than $4L$ points for which $\delta(q, v) > (1 + \epsilon)r$ but that map to the same bucket as q is at most 1/4. That completes the proof. □

5.2.1.2 Space and Time Complexity

The algorithm terminates after visiting at most $4L$ vectors in the candidate buckets. Given the configuration of Theorem 5.1, this means that the time complexity of the algorithm for query processing is $\mathcal{O}(dm^\rho)$, which is sub-linear in m.

As for space complexity of the algorithm, note that the index stores each data point L times. That implies the space required to build an LSH index has complexity $\mathcal{O}(mL) = \mathcal{O}(m^{1+\rho})$, which grows super-linearly with m. This growth rate can easily become prohibitive [Gionis et al., 1999, Buhler, 2001], particularly because it is often necessary to increase L to reach a higher accuracy, as the proof of Theorem 5.1 shows. How do we reduce this overhead and still obtain sub-linear query time? That is a question that has led to a flurry of research in the past.

One direction to address that question is to modify the search algorithm so that it visits multiple buckets from each of the L tables, instead of examining just a single bucket per table. That is the idea first explored by Panigrahy [2006]. In that work, the search algorithm is the same as in the standard version presented above, but in addition to searching the buckets for query q, it also performs many search operations for perturbed copies of q. While theoretically interesting, their method proves difficult to use in practice. That is because, the amount of noise needed to perturb a query depends on the distance of the nearest neighbor to q—a quantity that is unknown *a priori*.

Additionally, it is likely that a single bucket may be visited many times over as we invoke the search procedure on the copies of q.

Later, Lv et al. [2007] refined that theoretical result and presented a method that, instead of perturbing queries *randomly* and performing multiple hash computations and search invocations, utilizes a more efficient approach in deciding which buckets to probe within each table. In particular, their "multi-probe LSH" first finds the bucket associated with q, say $g_i(q)$. It then additionally visits other "adjacent" buckets where a bucket is adjacent if it is more likely to hold data points that are close to the vectors in $g_i(q)$.

The precise way their algorithm arrives at a set of adjacent buckets depends on the hash family itself. In their work, Lv et al. [2007] consider only a hash family for the Euclidean distance, and take advantage of the fact that adjacent buckets (which are in $[b]^\ell$) differ in each coordinate by at most 1—this becomes clearer when we review the LSH family for Euclidean distance in Section 5.3.3. This scheme was shown empirically to reduce by *an order of magnitude* the total number of hash tables that is required to achieve an accuracy greater than 0.9 on high-dimensional datasets.

Another direction is to improve the guarantees of the LSH family itself. As Theorem 5.1 indicates, $\rho = \log p_1 / \log p_2$ plays a critical role in the efficiency and effectiveness of the search algorithm, as well as the space complexity of the data structure. It makes sense, then, that improving ρ leads to smaller space overhead. Many works have explored advanced LSH families to do just that [Andoni and Indyk, 2008, Andoni et al., 2014, 2015]. We review some of these methods in more detail later in this chapter.

5.2.2 Back to the Approximate Retrieval Problem

A solution to PLEB of Definition 5.2 is a solution to ϵ-approximate top-k retrieval only if $r = \delta(q, u^*)$, where u^* is the k-th minimizer of $\delta(q, \cdot)$. But we do not know the minimal distance in advance! That begs the question: How does solving the PLEB problem help us solve the ϵ-approximate retrieval problem?

Indyk and Motwani [1998] argue that an efficient solution to this decision version of the problem leads directly to an efficient solution to the original problem. In effect, they show that ϵ-approximate retrieval can be reduced to PLEB. Let us review one simple, albeit inefficient reduction.

Let $\delta_* = \max_{u,v \in \mathcal{X}} \delta(u, v)$ and $\delta^* = \min_{u,v \in \mathcal{X}} \delta(u, v)$. Denote by Δ the aspect ratio: $\Delta = \delta_*/\delta^*$. Now, define a set of distances $\mathcal{R} = \{(1+\epsilon)^0, (1+\epsilon)^1, \ldots, \Delta\}$, and construct $|\mathcal{R}|$ LSH indices for each $r \in \mathcal{R}$.

Retrieving vectors for query q is a matter of performing binary search over $\mathcal{R}$ to find the minimal distance such that PLEB succeeds and returns a point $u \in \mathcal{X}$. That point u is the solution to the ϵ-approximate retrieval problem!

It is easy to see that such a reduction adds to the time complexity by a factor of $\mathcal{O}(\log\log_{1+\epsilon} \Delta)$, and to the space complexity by a factor of $\mathcal{O}(\log_{1+\epsilon} \Delta)$.

5.3 LSH Families

We have studied how LSH solves the PLEB problem of Definition 5.2, analyzed its time and space complexity, and reviewed how a solution to PLEB leads to a solution to the ϵ-approximate top-k retrieval problem of Definition 1.2. Throughout that discussion, we took for granted the existence of an LSH family that satisfies Definition 5.1 for a distance function of interest. In this section, we review example families and unpack their construction to complete the picture.

5.3.1 Hamming Distance

We start with the simpler case of Hamming distance over the space of binary vectors. That is, we assume that $\mathcal{X} \subset \{0,1\}^d$ and $\delta(u,v) = \|u-v\|_1$, measuring the number of coordinates in which the two vectors u and v differ. For this setup, a hash family that maps a vector to one of its coordinates at random—a technique that is also known as *bit sampling*—is an LSH family [Indyk and Motwani, 1998], as the claim below shows.

Theorem 5.2 *For $\mathcal{X} \subset \{0,1\}^d$ equipped with the Hamming distance, the family $\mathcal{H} = \{h_i \mid h_i(u) = u_i,\ 1 \leq i \leq d\}$ is $(r, (1+\epsilon)r, 1-r/d, 1-(1+\epsilon)r/d)$-sensitive.*

Proof. The proof is trivial. For a given r and two vectors $u, v \in \{0,1\}^d$, if $\|u-v\|_1 \leq r$, then $\mathbb{P}\Big[h_i(u) \neq h_i(v)\Big] \leq r/d$, so that $\mathbb{P}\Big[h_i(u) = h_i(v)\Big] \geq 1 - r/d$, and therefore $p_1 = 1 - r/d$. p_2 is derived similarly. □

5.3.2 Angular Distance

Consider next the angular distance between two real vectors $u, v \in \mathbb{R}^d$, defined as:

$$\delta(u,v) = \arccos\Big(\frac{\langle u, v\rangle}{\|u\|_2\|v\|_2}\Big). \tag{5.1}$$

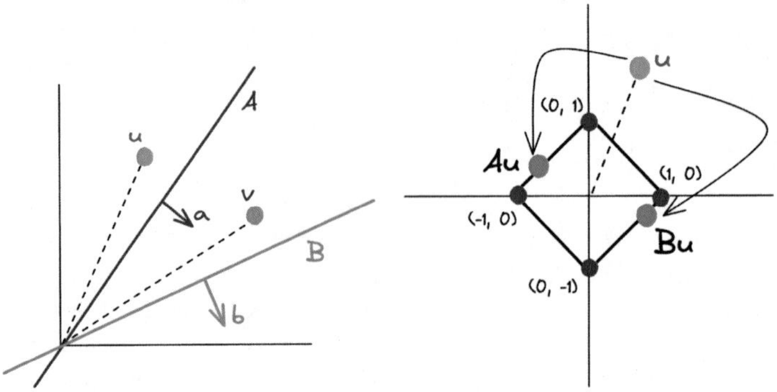

(a) Hyperplane LSH (b) Cross-polytope LSH

Fig. 5.2: Illustration of hyperplane and cross-polytope LSH functions for angular distance in $\mathbb{R}^2$. In hyperplane LSH, we draw random directions (a and b) to define hyperplanes (A and B), and record $+1$ or -1 depending on which side of the hyperplane a vector (u and v) lies. For example, $h_a(u) = -1$, $h_a(v) = +1$, and $h_b(u) = h_b(v) = -1$. It is easy to see that the probability of a hash collision for two vectors u and v correlates with the angle between them. A cross-polytope LSH function, on the other hand, randomly rotates and normalizes (using matrix A or B) the vector (u), and records the closest standard basis vector as its hash. Note that, the cross-polytope is the L_1 ball, which in $\mathbb{R}^2$ is a rotated square. As an example, $h_A(u) = -e_1$ and $h_B(u) = +e_1$.

5.3.2.1 Hyperplane LSH

For this distance function, one simple LSH family is the set of hash functions that project a vector onto a randomly chosen direction and record the sign of the projection. Put differently, a hash function in this family is characterized by a random hyperplane, which is in turn defined by a unit vector sampled uniformly at random. When applied to an input vector u, the function returns a binary value (from $\{-1, 1\}$) indicating on which side of the hyperplane u is located. This procedure, which is known as *sign random projections* or *hyperplane LSH* [Charikar, 2002], is illustrated in Figure 5.2(a) and formalized in the following claim.

Theorem 5.3 *For $\mathcal{X} \subset \mathbb{R}^d$ equipped with the angular distance of Equation (5.1), the family $\mathcal{H} = \{h_r \mid h_r(u) = \textsc{Sign}(\langle r, u\rangle),\ r \sim \mathbb{S}^{d-1}\}$ is $(\theta, (1+\epsilon)\theta, 1-\theta/\pi, 1-(1+\epsilon)\theta/\pi)$-sensitive for $\theta \in [0, \pi]$, and $\mathbb{S}^{d-1}$ denoting the d-dimensional hypersphere.*

Proof. If the angle between two vectors is θ, then the probability that a randomly chosen hyperplane lies between them is θ/π. As such, the probability that they lie on the same side of the hyperplane is $1 - \theta/\pi$. The claim follows. □

5.3.2.2 Cross-polytope LSH

There are a number of other hash families for the angular distance in addition to the basic construction above. *Spherical LSH* [Andoni et al., 2014] is one example, albeit a purely theoretical one—a single hash computation from that family alone is considerably more expensive than an exhaustive search over a million data points [Andoni et al., 2015]!

What is known as *Cross-polytope LSH* [Andoni et al., 2015, Terasawa and Tanaka, 2007] offers similar guarantees as the Spherical LSH but is a more practical construction. A function from this family randomly rotates an input vector first, then outputs the closest signed standard basis vector (e_i's for $1 \leq i \leq d$) as the hash value. This is illustrated for $\mathbb{R}^2$ in Figure 5.2(b), and stated formally in the following result.

Theorem 5.4 *For $\mathcal{X} \subset \mathbb{S}^{d-1}$ equipped with the angular distance of Equation (5.1) or equivalently the Euclidean distance, the following family constitutes an LSH:*

$$\mathcal{H} = \{h_R \mid h_R(u) = \underset{e \in \{\pm e_i\}_{i=1}^d}{\arg\min} \|e - \frac{Ru}{\|Ru\|_2}\|, R \in \mathbb{R}^{d \times d},\ R_{ij} \sim \mathcal{N}(0, 1)\},$$

where $\mathcal{N}(0,1)$ is the standard Gaussian distribution. The probability of collision for unit vectors $u, v \in \mathbb{S}^{d-1}$ with $\|u - v\| < \tau$ is:

$$\ln \frac{1}{\mathbb{P}\Big[h_R(u) = h_R(v)\Big]} = \frac{\tau^2}{4 - \tau^2} \ln d + \mathcal{O}_\tau\Big(\ln \ln d\Big).$$

Importantly:

$$\rho = \frac{\log p_1}{\log p_2} = \frac{1}{(1+\epsilon)^2} \frac{4 - (1+\epsilon)^2 r^2}{4 - r^2} + o(1).$$

Proof. We wish to show that, for two unit vectors $u, v \in \mathbb{S}^{d-1}$ with $\|u - v\| < \tau$, the expression above for the probability of a hash collision is correct. That, indeed, completes the proof of the theorem itself. To show that, we will take advantage of the spherical symmetry of Gaussian random variables—we used this property in the proof of Theorem 2.2.

By the spherical symmetry of Gaussians, without loss of generality, we can assume that $u = e_1$, the first standard basis, and $v = \alpha e_1 + \beta e_2$, where $\alpha^2 + \beta^2 = 1$ (so that v has unit norm) and $(\alpha - 1)^2 + \beta^2 = \tau^2$ (because the distance between u and v is τ).

Let us now model the collision probability as follows:

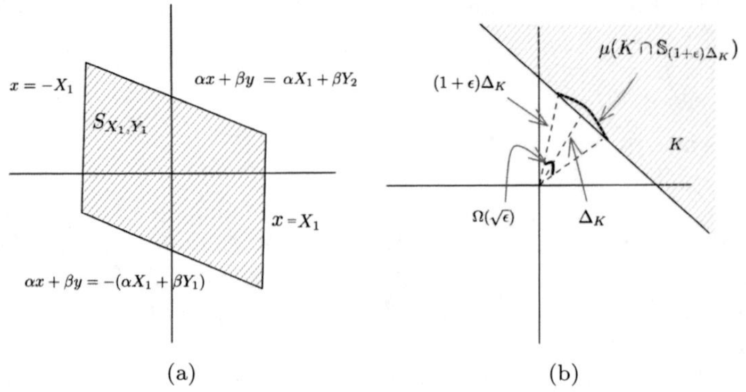

Fig. 5.3: Illustration of the set $S_{X_1,Y_1} = \{|x| \leq X_1 \wedge |\alpha x + \beta y| \leq \alpha X_1 + \beta Y_1\}$ in (a). Figure (b) visualizes the derivation of Equation (5.5).

$$\begin{aligned}
\mathbb{P}\Big[h(u) = h(v)\Big] &= 2d\,\mathbb{P}\Big[h(u) = h(v) = e_1\Big] \\
&= 2d \mathop{\mathbb{P}}_{X,Y\sim\mathcal{N}(0,I)} \Big[\forall\, i,\ |X_i| \leq X_1 \wedge |\alpha X_i + \beta Y_i| \leq \alpha X_1 + \beta Y_1\Big] \\
&= 2d \mathop{\mathbb{E}}_{X_1,Y_1\sim\mathcal{N}(0,1)} \left[\mathop{\mathbb{P}}_{X_2,Y_2} \Big[|X_2| \leq X_1 \wedge |\alpha X_2 + \beta Y_2| \leq \alpha X_1 + \beta Y_1\Big]^{d-1} \right].
\end{aligned} \tag{5.2}$$

The first equality is due again to the spherical symmetry of the hash functions and the fact that there are $2d$ signed standard basis vectors. The second equality simply uses the expressions for $u = e_1$ and $v = \alpha e_1 + \beta e_2$. The final equality follows because of the independence of the coordinates of X and Y, which are sampled from a d-dimensional isotropic Gaussian distribution.

The innermost term in Equation (5.2) is the Gaussian measure of the closed, convex set $\{|x| \leq X_1 \wedge |\alpha x + \beta y| \leq \alpha X_1 + \beta Y_1\}$, which is a bounded plane in $\mathbb{R}^2$. This set, which we denote by S_{X_1,Y_1}, is illustrated in Figure 5.3(a). Then we can expand Equation (5.2) as follows:

$$2d \mathop{\mathbb{E}}_{X_1,Y_1\sim\mathcal{N}(0,1)} \left[\mathop{\mathbb{P}}_{X_2,Y_2} \Big[S_{X_1,Y_1}\Big]^{d-1} \right] \tag{5.3}$$

$$= 2d \int_0^1 \mathop{\mathbb{P}}_{X_1,Y_1\sim\mathcal{N}(0,1)} \Big[\mathbb{P}[S_{X_1,Y_1}] \geq t^{\frac{1}{d-1}}\Big]\, dt. \tag{5.4}$$

We therefore need to expand $\mathbb{P}[S_{X_1,Y_1}]$ in order to complete the expression above. The rest of the proof derives that quantity.

Step 1. Consider $\mathbb{P}[S_{X_1,Y_1}] = \mathcal{G}(S_{X_1,Y_1})$, which is the standard Gaussian measure of the set S_{X_1,Y_1}. In effect, we are interested in $\mathcal{G}(S)$ for some bounded convex subset $S \subset \mathbb{R}^2$. We need the following lemma to derive an expression for $\mathcal{G}(S)$. But first define $\mu_A(r)$ as the Lebesgue measure of the intersection of a circle of radius r ($\mathbb{S}_r$) with the set A, normalized by the circumference of $\mathbb{S}_r$, so that $0 \leq \mu_A(r) \leq 1$ is a probability measure:

$$\mu_A(r) \triangleq \frac{\mu(A \cap \mathbb{S}_r)}{2\pi r},$$

and denote by Δ_A the distance from the origin to A (i.e., $\Delta_A \triangleq \inf\{r > 0 \mid \mu_A(r) > 0\}$).

Lemma 5.1 *For the closed set $A \subset \mathbb{R}^2$ with $\mu_A(r)$ non-decreasing:*

$$\sup_{r>0} \left(\mu_A(r) \cdot e^{-r^2/2}\right) \leq \mathcal{G}(A) \leq e^{-\Delta_A^2/2}.$$

Proof. The upper-bound can be derived as follows:

$$\mathcal{G}(A) = \int_0^\infty r\mu_A(r) \cdot e^{-r^2/2} dr \leq \int_{\Delta_A}^\infty r e^{-r^2/2} dr = e^{-\Delta_A^2/2}.$$

For the lower-bound:

$$\mathcal{G}(A) = \int_0^\infty r\mu_A(r) \cdot e^{-r^2/2} dr \geq \mu_A(r') \int_{r'}^\infty r e^{-r^2/2} dr = \mu_A(r') e^{-(r')^2/2},$$

for all $r' > 0$. The inequality holds because $\mu_A(\cdot)$ is non-decreasing. □

Now, $K^{\complement} \triangleq S_{X_1,Y_1}$ is a convex set, so for its complement, $K \subset \mathbb{R}^2$, $\mu_K(\cdot)$ is non-decreasing. Using the above lemma, that fact implies the following for small ϵ:

$$\Omega(\sqrt{\epsilon} \cdot e^{-(1+\epsilon)^2 \Delta_K^2/2}) \leq \mathcal{G}(K) \leq e^{-\Delta_K^2/2}.$$

The lower-bound uses the fact that $\mu_K\Big((1+\epsilon)\Delta_K\Big) = \Omega(\sqrt{\epsilon})$, because:

$$\mu(K \cap \mathbb{S}_{(1+\epsilon)\Delta_K}) = (1+\epsilon)\Delta_K \arccos\left(\frac{\Delta_K}{(1+\epsilon)\Delta_K}\right) \approx (1+\epsilon)\Delta_K\sqrt{\epsilon}. \quad (5.5)$$

See Figure 5.3(b) for a helpful illustration.

Since we are interested in the measure of $K^{\complement} = S_{X_1,Y_1}$, we can apply the result above directly to obtain:

$$1 - e^{-\Delta(u,v)^2/2} \leq \mathbb{P}[S_{X_1,Y_1}] \leq 1 - \Omega\Big(\sqrt{\epsilon} \cdot e^{-(1+\epsilon)^2 \Delta(u,v)^2/2}\Big), \quad (5.6)$$

where we use the notation $\Delta_K = \Delta(u,v) = \min\{u, \alpha u + \beta v\}$.

Step 2. For simplicity, first consider the side of Equation (5.6) that does not depend on ϵ, and substitute that into Equation (5.3). We obtain:

$$\begin{aligned}2d\int_0^1 \mathop{\mathbb{P}}_{X_1,Y_1\sim\mathcal{N}(0,1)}&\left[\mathbb{P}[S_{X_1,Y_1}]\geq t^{\frac{1}{d-1}}\right]dt\\
&= 2d\int_0^1 \mathop{\mathbb{P}}_{X_1,Y_1\sim\mathcal{N}(0,1)}\left[e^{-\Delta(X_1,Y_1)^2/2}\leq 1-t^{\frac{1}{d-1}}\right]dt\\
&= 2d\int_0^1 \mathop{\mathbb{P}}_{X_1,Y_1\sim\mathcal{N}(0,1)}\left[\Delta(X_1,Y_1)\geq\sqrt{-2\log\left(1-t^{\frac{1}{d-1}}\right)}\right]dt.\end{aligned} \tag{5.7}$$

Step 3. We are left with bounding $\mathbb{P}[\Delta(X_1,Y_1)\geq\theta]$. $\Delta(X_1,Y_1)\geq\theta$ is, by definition, the set that is the intersection of two half-planes: $X_1\geq\theta$ and $\alpha X_1+\beta Y_1\geq\theta$. If we denote this set by K, then we are again interested in the Gaussian measure of K. For small ϵ, we can apply the lemma above to show that:

$$\Omega\left(\epsilon e^{-(1+\epsilon)^2\Delta_K^2}\right)\leq\mathcal{G}(K)\leq e^{-\Delta_K^2/2}, \tag{5.8}$$

where the constant factor in Ω depends on the angle between the two half-planes. That is because $\mu(K\cap\mathbb{S}_{(1+\epsilon)\Delta_K})$ is ϵ times that angle.

It is easy to see that $\Delta_K^2=\frac{4}{4-\tau^2}\cdot\theta^2$, so that we arrive at the following for small ϵ and every $\theta\geq 0$:

$$\Omega_\tau\left(\epsilon\cdot e^{-(1+\epsilon)^2\cdot\frac{4}{4-\tau^2}\cdot\frac{\theta^2}{2}}\right)\leq\mathop{\mathbb{P}}_{X_1,Y_1\sim\mathcal{N}(0,1)}\left[\Delta(X_1,Y_1)\geq\theta\right]\leq e^{-\frac{4}{4-\tau^2}\cdot\frac{\theta^2}{2}}. \tag{5.9}$$

Step 4. Substituting Equation (5.9) into Equation (5.7) yields:

$$\begin{aligned}2d\int_0^1 \mathop{\mathbb{P}}_{X_1,Y_1\sim\mathcal{N}(0,1)}&\left[\Delta(X_1,Y_1)\geq\sqrt{-2\log\left(1-t^{\frac{1}{d-1}}\right)}\right]dt\\
&= 2d\int_0^1\left(1-t^{\frac{1}{d-1}}\right)^{\frac{4}{4-\tau^2}}dt\\
&= 2d(d-1)\int_0^1(1-x)^{\frac{4}{4-\tau^2}}x^{d-2}dt\\
&= 2d(d-1)B\left(\frac{8-\tau^2}{4-\tau^2};d-1\right)\\
&= 2d\Theta_\tau(1)d^{-\frac{4}{4-\tau^2}},\end{aligned}$$

where B denotes the Beta function and the last step uses the Stirling approximation.

The result above can be expressed as follows:

$$\ln\frac{1}{\mathbb{P}[h(u)=h(v)]}=\frac{\tau^2}{4-\tau^2}\ln d\pm\mathcal{O}_\tau(1).$$

Step 5. Repeating Steps 2 through 4 with the expressions that involve ϵ in Equations (5.6) and (5.9) gives the desired result. □

Finally, Andoni et al. [2015] show that, instead of applying a random rotation using Gaussian random variables, it is sufficient to use a pseudo-random rotation based on Fast Hadamard Transform. In effect, they replace the random Gaussian matrix R in the construction above with three consecutive applications of HD, where H is the Hadamard matrix and D is a random diagonal sign matrix (where the entries on the diagonal take values from $\{\pm 1\}$).

5.3.3 Euclidean Distance

Datar et al. [2004] proposed the first LSH family for the Euclidean distance, $\delta(u, v) = \|u - v\|_2$. Their construction relies on the notion of *p-stable distributions* which we define first.

Definition 5.3 (p-stable Distribution) A distribution $\mathcal{D}_p$ is said to be p-stable if $\sum_{i=1}^{n} \alpha_i Z_i$, where $\alpha_i \in \mathbb{R}$ and $Z_i \sim \mathcal{D}_p$, has the same distribution as $\|\alpha\|_p Z$, where $\alpha = [\alpha_1, \alpha_2, \ldots, \alpha_n]$ and $Z \sim \mathcal{D}_p$. As an example, the Gaussian distribution is 2-stable.

Let us state this property slightly differently so it is easier to understand its connection to LSH. Suppose we have an arbitrary vector $u \in \mathbb{R}^d$. If we construct a d-dimensional random vector α whose coordinates are independently sampled from a p-stable distribution $\mathcal{D}_p$, then the inner product $\langle \alpha, u \rangle$ is distributed according to $\|u\|_p Z$ where $Z \sim \mathcal{D}_p$. By linearity of inner product, we can also see that $\langle \alpha, u \rangle - \langle \alpha, v \rangle$, for two vectors $u, v \in \mathbb{R}^d$, is distributed as $\|u - v\|_p Z$. This particular fact plays an important role in the proof of the following result.

Theorem 5.5 *For $\mathcal{X} \subset \mathbb{R}^d$ equipped with the Euclidean distance, a 2-stable distribution $\mathcal{D}_2$, and the uniform distribution U over the interval $[0, r]$, the following family is $(r, (1+\epsilon)r, p(r), p((1+\epsilon)r))$-sensitive:*

$$\mathcal{H} = \{h_{\alpha,\beta} \mid h_{\alpha,\beta}(u) = \lfloor \frac{\langle \alpha, u \rangle + \beta}{r} \rfloor, \ \alpha \in \mathbb{R}^d, \ \alpha_i \sim \mathcal{D}_2, \ \beta \sim U[0, r]\},$$

where:

$$p(x) = \int_{t=0}^{r} \frac{1}{x} f\Big(\frac{t}{x}\Big)\Big(1 - \frac{t}{r}\Big) dt,$$

and f is the probability density function of the absolute value *of $\mathcal{D}_2$.*

Proof. The key to proving the claim is modeling the probability of a hash collision for two arbitrary vectors u and v: $\mathbb{P}\Big[h_{\alpha,\beta}(u) = h_{\alpha,\beta}(v)\Big]$. That event can be expressed as follows:

$$\mathbb{P}\Big[h_{\alpha,\beta}(u) = h_{\alpha,\beta}(v)\Big] = \mathbb{P}\Big[\Big\lfloor\frac{\langle\alpha, u\rangle + \beta}{r}\Big\rfloor = \Big\lfloor\frac{\langle\alpha, v\rangle + \beta}{r}\Big\rfloor\Big]$$
$$= \mathbb{P}\Big[\underbrace{|\langle\alpha, u - v\rangle| < r}_{Event\ A} \wedge$$
$$\underbrace{\langle\alpha, u\rangle + \beta \textit{ and } \langle\alpha, v\rangle + \beta \textit{ do not straddle an integer}}_{Event\ B}\Big].$$

Using the 2-stability of α, Event A is equivalent to $\|u - v\|_2|Z| < r$, where Z is drawn from $\mathcal{D}_2$. The probability of the complement of Event B is simply the ratio between $\langle\alpha, u - v\rangle$ and r. Putting all that together, we obtain that:

$$\mathbb{P}\Big[h_{\alpha,\beta}(u) = h_{\alpha,\beta}(v)\Big] = \int_{z=0}^{\frac{r}{\|u-v\|_2}} f(z)\Big(1 - \frac{z\|u - v\|_2}{r}\Big)dz$$
$$= \int_{t=0}^{r} \frac{1}{\|u - v\|_2} f(\frac{t}{\|u - v\|_2})\Big(1 - \frac{t}{r}\Big)dt,$$

where we derived the last equality by the variable change $t = z\|u - v\|_2$. Therefore, if $\|u - v\| \leq x$:

$$\mathbb{P}\Big[h_{\alpha,\beta}(u) = h_{\alpha,\beta}(v)\Big] \geq \int_{t=0}^{r} \frac{1}{x} f(\frac{t}{x})\Big(1 - \frac{t}{r}\Big)dt = p(x).$$

It is easy to complete the proof from here. □

5.3.4 Inner Product

Many of the arguments that establish the existence of an LHS family for a distance function of interest rely on triangle inequality. Inner product as a measure of similarity, however, does not enjoy that property. As such, developing an LSH family for inner product requires that we somehow transform the problem from MIPS to NN search or MCS search, as was the case in Chapter 4.

Finding the right transformation that results in improved hash quality—as determined by ρ—is the question that has been explored by several works in the past [Neyshabur and Srebro, 2015, Shrivastava and Li, 2015, 2014, Yan et al., 2018].

Let us present a simple example. Note that, we may safely assume that queries are unit vectors (i.e., $q \in \mathbb{S}^{d-1}$), because the norm of the query does not change the outcome of MIPS.

Now, define the transformation $\phi_d : \mathbb{R}^d \rightarrow \mathbb{R}^{d+1}$, first considered by Bachrach et al. [2014], as follows: $\phi_d(u) = [u, \sqrt{1 - \|u\|_2^2}]$. Apply this transformation to data points in $\mathcal{X}$. Clearly, $\|\phi_d(u)\|_2 = 1$ for all $u \in \mathcal{X}$. Separately, pad the query points with a single 0: $\phi_q(v) = [v; 0] \in \mathbb{R}^{d+1}$.

We can immediately verify that $\langle q, u \rangle = \langle \phi_q(q), \phi_d(u) \rangle$ for a query q and data point u. But by applying the transformations $\phi_d(\cdot)$ and $\phi_q(\cdot)$, we have reduced the problem to MCS! As such, we may use any of existing LSH families that we have seen for angular distance in Section 5.3.2 for MIPS.

There has been much debate over the suitability of the standard LSH framework for inner product, with some works extending the framework to what is known as *asymmetric* LSH [Shrivastava and Li, 2014, 2015]. It turns out, however, that none of that is necessary. In fact, as Neyshabur and Srebro [2015] argued formally and demonstrated empirically, the simple scheme we described above sufficiently addresses MIPS.

5.4 Closing Remarks

Much like branch-and-bound algorithms, an LSH approach to top-k retrieval rests on a solid theoretical foundation. There is a direct link between all that is developed theoretically and the accuracy of an LSH-based top-k retrieval system.

Like tree indices, too, the LSH literature is arguably mature. There is therefore not a great deal of open questions left to investigate in its foundation, with many recent works instead exploring learnt hash functions or its applications in other domains.

What remains open and exciting in the context of top-k retrieval, however, is the possibility of extending the theory of LSH to explain the success of other retrieval algorithms. We will return to this discussion in Chapter 7.

References

A. Andoni and P. Indyk. Near-optimal hashing algorithms for approximate nearest neighbor in high dimensions. *Communications of the ACM*, 51(1): 117–122, 1 2008.

A. Andoni, P. Indyk, H. L. Nguyen, and I. Razenshteyn. Beyond locality-sensitive hashing. In *Proceedings of the 2014 Annual ACM-SIAM Symposium on Discrete Algorithms*, pages 1018–1028, 2014.

A. Andoni, P. Indyk, T. Laarhoven, I. Razenshteyn, and L. Schmidt. Practical and optimal lsh for angular distance. In *Proceedings of the 28th International Conference on Neural Information Processing Systems - Volume 1*, pages 1225–1233, 2015.

Y. Bachrach, Y. Finkelstein, R. Gilad-Bachrach, L. Katzir, N. Koenigstein, N. Nice, and U. Paquet. Speeding up the xbox recommender system using a euclidean transformation for inner-product spaces. In *Proceedings of the 8th ACM Conference on Recommender Systems*, page 257–264, 2014.

J. Buhler. Efficient large-scale sequence comparison by locality-sensitive hashing. *Bioinformatics*, 17(5):419–428, 05 2001.

M. S. Charikar. Similarity estimation techniques from rounding algorithms. In *Proceedings of the Thiry-Fourth Annual ACM Symposium on Theory of Computing*, pages 380–388, 2002.

M. Datar, N. Immorlica, P. Indyk, and V. S. Mirrokni. Locality-sensitive hashing scheme based on p-stable distributions. In *Proceedings of the 20th Annual Symposium on Computational Geometry*, pages 253–262, 2004.

A. Gionis, P. Indyk, and R. Motwani. Similarity search in high dimensions via hashing. In *Proceedings of the 25th International Conference on Very Large Data Bases*, pages 518–529, 1999.

P. Indyk and R. Motwani. Approximate nearest neighbors: Towards removing the curse of dimensionality. In *Proceedings of the 30th Annual ACM Symposium on Theory of Computing*, pages 604–613, 1998.

Q. Lv, W. Josephson, Z. Wang, M. Charikar, and K. Li. Multi-probe lsh: Efficient indexing for high-dimensional similarity search. In *Proceedings of the 33rd International Conference on Very Large Data Bases*, pages 950–961, 2007.

B. Neyshabur and N. Srebro. On symmetric and asymmetric lshs for inner product search. In *Proceedings of the 32nd International Conference on International Conference on Machine Learning - Volume 37*, pages 1926–1934, 2015.

R. Panigrahy. Entropy based nearest neighbor search in high dimensions. In *Proceedings of the Seventeenth Annual ACM-SIAM Symposium on Discrete Algorithm*, pages 1186–1195, 2006.

A. Shrivastava and P. Li. Asymmetric lsh (alsh) for sublinear time maximum inner product search (mips). In *Proceedings of the 27th International Conference on Neural Information Processing Systems - Volume 2*, pages 2321–2329, 2014.

A. Shrivastava and P. Li. Improved asymmetric locality sensitive hashing (alsh) for maximum inner product search (mips). In *Proceedings of the Thirty-First Conference on Uncertainty in Artificial Intelligence*, pages 812–821, 2015.

K. Terasawa and Y. Tanaka. Spherical lsh for approximate nearest neighbor search on unit hypersphere. In *Proceedings of the 10th International Conference on Algorithms and Data Structures*, pages 27–38, 2007.

X. Yan, J. Li, X. Dai, H. Chen, and J. Cheng. Norm-ranging lsh for maximum inner product search. In *Proceedings of the 32nd International Conference on Neural Information Processing Systems*, pages 2956–2965, 2018.

Chapter 6
Graph Algorithms

Abstract We have seen two major classes of algorithms that approach the top-k retrieval problem in their own unique ways. One recursively partitions a vector collection to model its geometry, and the other hashes the vectors into predefined buckets to reduce the search space. Our next class of algorithms takes yet a different view of the question. At a high level, our third approach is to "walk" through a collection, hopping from one vector to another, where every hop gets us *spatially* closer to the optimal solution. This chapter reviews algorithms that use a graph data structure to implement that idea.

6.1 Intuition

The most natural way to understand a spatial walk through a collection of vectors is by casting it as traversing a (directed) connected graph. As we will see, whether the graph is directed or not depends on the specific algorithm itself. But the graph must regardless be *connected*, so that there always exists at least one path between every pair of nodes. This ensures that we can walk through the graph no matter where we begin our traversal.

Let us write $G(\mathcal{V}, \mathcal{E})$ to refer to such a graph, whose set of *vertices* or *nodes* are denoted by $\mathcal{V}$, and its set of edges by $\mathcal{E}$. So, for $u, v \in \mathcal{V}$ in a directed graph, if $(u, v) \in \mathcal{E}$, we may freely move from node u to node v. Hopping from v to u is not possible if $(v, u) \notin \mathcal{E}$. Because we often need to talk about the set of nodes that can be reached by a single hop from a node u—known as the neighbors of u—we give it a special symbol and define that set as follows: $N(u) = \{v \mid (u, v) \in \mathcal{E}\}$.

The idea behind the algorithms in this chapter is to construct a graph in the pre-processing phase and use that as an index of a vector collection for top-k retrieval. To do that, we must decide what is a node in the graph (i.e., define the set $\mathcal{V}$), how nodes are linked to each other ($\mathcal{E}$), and, importantly, what the search algorithm looks like.

S. Bruch, *Foundations of Vector Retrieval*, https://doi.org/10.1007/978-3-031-55182-6_6

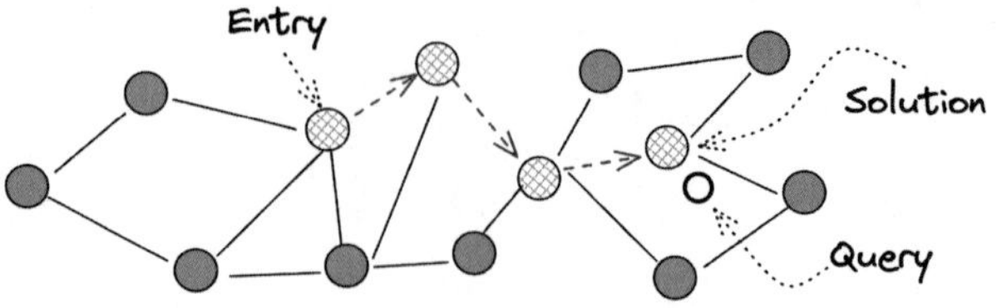

Fig. 6.1: Illustration of the greedy traversal algorithm for finding the top-1 solution on an example (undirected) graph. The procedure enters the graph from an arbitrary "entry" node. It then compares the distance of the node to query q with the distance of its neighbors to q, and either terminates if no neighbor is closer to q than the node itself, or advances to the closest neighbor. It repeats this procedure until the terminal condition is met. The research question in this chapter concerns the construction of the edge set: How do we construct a sparse graph which can be traversed greedily while providing guarantees on the (near-)optimality of the solution

The set of nodes $\mathcal{V}$ is easy to construct: Simply designate every vector in the collection $\mathcal{X}$ as a unique node in G, so that $|\mathcal{X}| = |\mathcal{V}|$. There should, therefore, be no ambiguity if we referred to a node as a vector. We use both terms interchangeably.

What properties should the edge set $\mathcal{E}$ have? To get a sense of what is required of the edge set, it would help to consider the search algorithm first. Suppose we are searching for the top-1 vector closest to query q, and assume that we are, at the moment, at an arbitrary node u in G.

From node u, we can have a look around and assess if any of our neighbors in $N(u)$ is closer to q. By doing so, we find ourselves in one of two situations. Either we encounter no such neighbor, so that u has the smallest distance to q among its neighbors. If that happens, ideally, we want u to also have the smallest distance to q among *all* vectors. In other words, in an ideal graph, a local optimum coincides with the global optimum.

Alternatively, we may find one such neighbor $v \in N(u)$ for which $\delta(q, v) < \delta(q, u)$ and $v = \arg\min_{w \in N(u)} \delta(q, w)$. In that case, the ideal graph is one where the following event takes place: If we moved from u to v, and repeated the process above in the context of $N(v)$ and so on, we will ultimately arrive at a local optimum (which, by the previous condition, is the global optimum). Terminating the algorithm then would therefore give us the optimal solution to the top-1 retrieval problem.

Put differently, in an ideal graph, if moving from a node to any of its neighbors does not get us spatially closer to q, it is because the current node is the optimal solution to the top-1 retrieval problem for q.

Algorithm 3: Greedy search algorithm for top-k retrieval over a graph index.

Input: Graph $G = (\mathcal{V}, \mathcal{E})$ over collection $\mathcal{X}$ with distance $\delta(\cdot, \cdot)$; query point q; entry node $s \in \mathcal{V}$; retrieval depth k.
Result: Exact top-k solution for q.

$Q \leftarrow \{s\}$; ▷ Q is a priority queue
while Q changed in the previous iteration **do**
  $\mathcal{S} \leftarrow \bigcup_{u \in Q} N(u)$
  $v \leftarrow \arg\min_{u \in \mathcal{S}} \delta(q, u)$
  Q.Insert(v)
  if $|Q| \geq k$ **then**
    Q.Pop() ; ▷ Removes the node with the largest distance to q
  end if
end while
return Q

On a graph with that property, the procedure of starting from any node in the graph, hopping to a neighbor that is closer to q, and repeating this procedure until no such neighbor exists, gives the optimal solution. That procedure is the familiar *best-first-search* algorithm, which we illustrate on a toy graph in Figure 6.1. That will be our base search algorithm for top-1 retrieval.

Extending the search algorithm to top-k requires a minor modification to the procedure above. It begins by initializing a priority queue of size k. When we visit a new node, we add it to the queue if its distance with q is smaller than the minimum distance among the nodes already in the queue. We keep moving from a node in the queue to its neighbors until the queue stabilizes (i.e., no unseen neighbor of any of the nodes in the queue has a smaller distance to q). This is described in Algorithm 3.

Note that, assuming $\delta(\cdot, \cdot)$ is proper, it is easy to see that the top-1 optimality guarantee immediately implies top-k optimality—you should verify this claim as an exercise. It therefore suffices to state our requirements in terms of top-1 optimality alone. So, ideally, $\mathcal{E}$ should guarantee that traversing G in a best-first-search manner yields the optimal top-1 solution.

6.1.1 The Research Question

It is trivial to construct an edge set that provides the desired optimality guarantee: Simply add an edge between every pair of nodes, completing the graph! The greedy search algorithm described above will take us to the optimal solution.

However, such a graph not only has high space complexity, but it also has a linear query time complexity. That is because, the very first step (which also happens to be the last step) involves comparing the distance of q to the entry node, with the distance of q to every other node in the graph! We are better off exhaustively scanning the entire collection in a flat index.

The research question that prompted the algorithms we are about to study in this chapter is whether there exists a relatively *sparse* graph that has the optimality guarantee we seek or that can instead provide guarantees for the more relaxed, ϵ-approximate top-k retrieval problem.

As we will learn shortly, with a few notable exceptions, all constructions of $\mathcal{E}$ proposed thus far in the literature for high-dimensional vectors amount to heuristics that attempt to *approximate* a theoretical graph but come with no guarantees. In fact, in almost all cases, their worst-case complexity is no better than exhaustive search. Despite that, many of these heuristics work remarkably well in practice on real datasets, making graph-based methods one of the most widely adopted solutions to the approximate top-k retrieval problem.

In the remainder of this chapter, we will see classes of theoretical graphs that were developed in adjacent scientific disciplines, but that are seemingly suitable for the (approximate) top-k retrieval problem. As we introduce these graphs, we also examine representative algorithms that aim to build an approximation of such graphs in high dimensions, and review their properties.

We note, however, that the literature on graph-based methods is vast and growing still. There is a plethora of studies that experiment with (minor or major) adjustments to the basic idea described earlier, or that empirically compare and contrast different algorithmic flavors on real-world datasets. This chapter does not claim to, nor does it intend to cover the explosion of material on graph-based algorithms. Instead, it limits its scope to the foundational principles and ground-breaking works that are theoretically somewhat interesting. We refer the reader to existing reports and surveys for the full spectrum of works on this topic [Wang et al., 2021, Li et al., 2020].

6.2 The Delaunay Graph

One classical graph that satisfies the conditions we seek and guarantees the optimality of the solution obtained by best-first-search traversal is the Delaunay graph [Delaunay, 1934, Fortune, 1997]. It is easier to understand the construction of the Delaunay graph if we consider instead its dual: the Voronoi diagram. So we begin with a description of the Voronoi diagram and Voronoi regions.

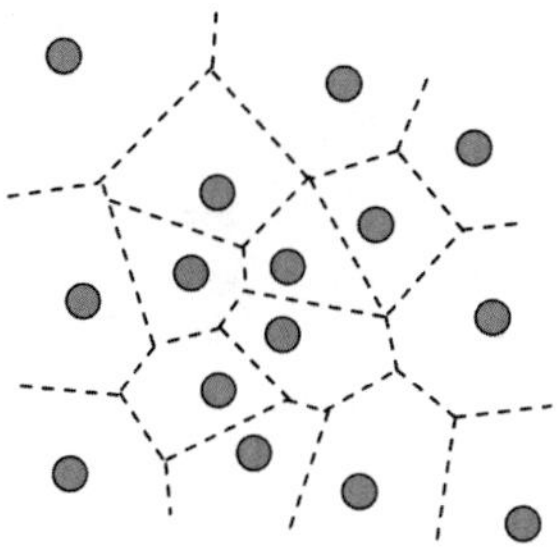

(a) Voronoi diagram

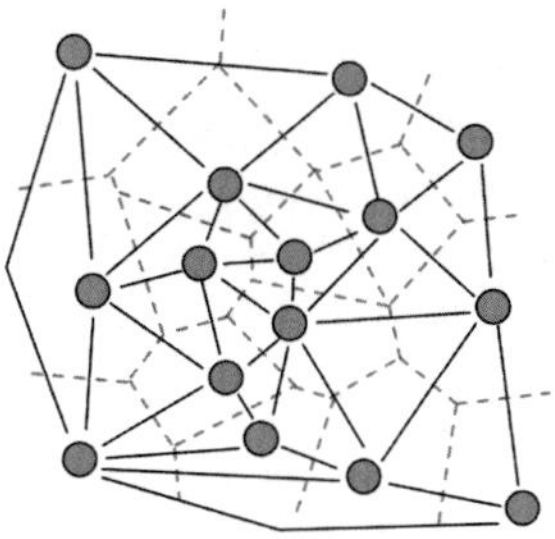

(b) Delaunay graph

Fig. 6.2: Visualization of the Voronoi diagram (a) and its dual, the Delaunay graph (b) for an example collection $\mathcal{X}$ of points in $\mathbb{R}^2$. A Voronoi region associated with a point u (shown here as the area contained within the dashed lines) is a set of points whose nearest neighbor in $\mathcal{X}$ is u. The Delaunay graph is an undirected graph whose nodes are points in $\mathcal{X}$ and two nodes are connected (shown as solid lines) if their Voronoi regions have a non-empty intersection.

6.2.1 Voronoi Diagram

For the moment, suppose δ is the Euclidean distance and that we are in $\mathbb{R}^2$. Suppose further that we have a collection $\mathcal{X}$ of just two points u and v on the plane. Consider now the subset of $\mathbb{R}^2$ comprising of all the points to which u is the closest point from $\mathcal{X}$. Similarity, we can identify the subset to which v is the closest point. These two subsets are, in fact, partitions of the plane and are separated by a line—the points on this line are equidistant to u and v. In other words, two points in $\mathbb{R}^2$ induce a partitioning of the plane where each partition is "owned" by a point and describes the set of points that are closer to it than they are to the other point.

We can trivially generalize that notion to more than two points and, indeed, to higher dimensions. A collection $\mathcal{X}$ of points in $\mathbb{R}^d$ partitions the space into unique regions $\mathcal{R} = \bigcup_{u \in \mathcal{X}} \mathcal{R}_u$, where the region $\mathcal{R}_u$ is owned by point $u \in \mathcal{X}$ and represents the set of points to which u is the closest point in $\mathcal{X}$. Formally, $\mathcal{R}_u = \{x \mid u = \arg\min_{v \in \mathcal{X}} \delta(x, v)\}$. Note that, each region is a convex polytope that is the intersection of half-spaces. The set of regions is known as the Voronoi diagram for the collection $\mathcal{X}$ and is illustrated in Figure 6.2(a) for an example collection in $\mathbb{R}^2$.

6.2.2 Delaunay Graph

The Delaunay graph for $\mathcal{X}$ is, in effect, a graph representation of its Voronoi diagram. The nodes of the graph are trivially the points in $\mathcal{X}$, as before. We place an edge between two nodes u and v if their Voronoi regions have a non-empty intersection: $\mathcal{R}_u \cap \mathcal{R}_v \neq \emptyset$. Clearly, by construction, this graph is undirected. An example of this graph is rendered in Figure 6.2(b).

There is an important technical detail that is worth noting. The Delaunay graph for a collection $\mathcal{X}$ is unique if the points in $\mathcal{X}$ are in *general position* [Fortune, 1997]. A collection of points are said to be in general position if the following two conditions are satisfied. First, no n points from $\mathcal{X} \subset \mathbb{R}^d$, for $2 \leq n \leq d+1$, must lie on a $(n-2)$-flat. Second, no $n+1$ points must lie on any $(n-2)$-dimensional hypersphere. In $\mathbb{R}^2$, as an example, for a collection of points to be in general position, no three points may be co-linear, and no four points co-circular.

We must add that, the detail above is generally satisfied in practice. Importantly, if the vectors in our collection are independent and identically distributed, then the collection is almost surely in general position. That is why we often take that technicality for granted. So from now on, we assume that the Delaunay graph of a collection of points is unique.

6.2.3 Top-1 Retrieval

We can immediately recognize the importance of Voronoi regions: They geometrically capture the set of queries for which a point from the collection is the solution to the top-1 retrieval problem. But what is the significance of the dual representation of this geometrical concept? How does the Delaunay graph help us solve the top-1 retrieval problem?

For one, the Delaunay graph is a compact representation of the Voronoi diagram. Instead of describing polytopes, we need only to record edges between neighboring nodes. But, more crucially, as the following claim shows, we can traverse the Delaunay graph greedily, and reach the optimal top-1 solution from any node. In other words, the Delaunay graph has the desired property we described in Section 6.1.

Theorem 6.1 *Let $G = (\mathcal{V}, \mathcal{E})$ be a graph that contains the Delaunay graph of m vectors $\mathcal{X} \subset \mathbb{R}^d$. The best-first-search algorithm over G gives the optimal solution to the top-1 retrieval problem for any arbitrary query q if $\delta(\cdot, \cdot)$ is proper.*

The proof of the result above relies on an important property of the Delaunay graph, which we state first.

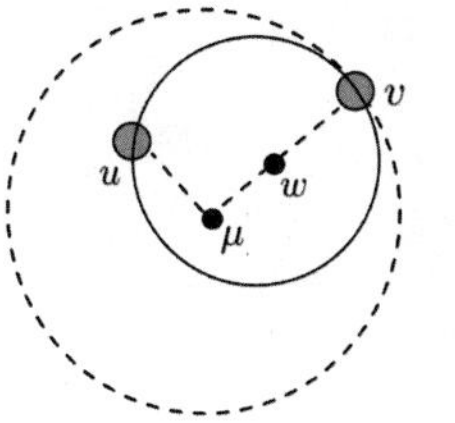

Fig. 6.3: Illustration of the second case in the proof of Lemma 6.1.

Lemma 6.1 *Let $G = (\mathcal{V}, \mathcal{E})$ be the Delaunay graph of a collection of points $\mathcal{X} \subset \mathbb{R}^d$, and let B be a ball centered at μ that contains two points $u, v \in \mathcal{X}$, with radius $r = \min(\delta(\mu, u), \delta(\mu, v))$, for a continuous and proper distance function $\delta(\cdot, \cdot)$. Then either $(u, v) \in \mathcal{E}$ or there exists a third point in $\mathcal{X}$ that is contained in B.*

Proof. Suppose there is no other point in $\mathcal{X}$ that is contained in B. We must show that, in that case, $(u, v) \in \mathcal{E}$.

There are two cases. The first and easy case is when u and v are on the surface of B. Clearly, u and v are equidistant from μ. Because there are no other points in B, we can conclude that μ lies in the intersection of $\mathcal{R}_u$ and $\mathcal{R}_v$, the Voronoi regions associated with u and v. That implies $(u, v) \in \mathcal{E}$.

In the second case, suppose $\delta(\mu, u) < \delta(\mu, v)$, so that v is on the surface of B and u is in its interior. Consider the function $f(\omega) = \delta(v, \omega) - \delta(u, \omega)$. Clearly, $f(v) < 0$ and $f(\mu) > 0$. Therefore, there must be a point $w \in B$ on the line segment $\mu + \lambda(v - \mu)$ for $\lambda \in [0, 1]$ for which $f(w) = 0$. That implies that $\delta(w, u) = \delta(w, v)$. Furthermore, v is the closest point on the surface of B to w, so that the ball centered at w with radius $\delta(w, v)$ is entirely contained in B. This is illustrated in Figure 6.3.

Importantly, no other point in $\mathcal{X}$ is closer to w than u and v. So w rests in the intersection of $\mathcal{R}_u$ and $\mathcal{R}_v$, and $(u, v) \in \mathcal{E}$. □

Proof of Theorem 6.1. We prove the result for the case where δ is the Euclidean distance and leave the proof of the more general case as an exercise. (Hint: To prove the general case you should make the line segment argument as in the proof of Lemma 6.1.)

Suppose the greedy search for q stops at some local optimum u that is different from the global optimum, u^*, and that $(u, u^*) \notin \mathcal{E}$—otherwise, the algorithm must terminate at u^* instead. Let $r = \delta(q, u)$.

By assumption we have that the ball centered at q with radius r, $B(q, r)$, is non-empty because it must contain u^* whose distance to q is less than r. Let v be the point in this ball that is closest to u. Consider now the ball $B((u + v)/2, \delta(u, v)/2)$. This ball is empty: otherwise v would not be the closest point to u. By Lemma 6.1, we must have that $(u, v) \in \mathcal{E}$. This is a contradiction because the greedy search cannot stop at u. □

Notice that Theorem 6.1 holds for any graph that *contains* the Delaunay graph. The next theorem strengthens this result to show that the Delaunay graph represents the minimal edge set that guarantees an optimal solution through greedy traversal.

Theorem 6.2 *The Delaunay graph is the minimal graph over which the best-first-search algorithm gives the optimal solution to the top-*1 *retrieval problem.*

In other words, if a graph does not contain the Delaunay graph, then we can find queries for which the greedy traversal from an entry point does not produce the optimal top-1 solution.

Proof of Theorem 6.2. Suppose that the data points $\mathcal{X}$ are in general position. Suppose further that $G = (\mathcal{V}, \mathcal{E})$ is a graph built from $\mathcal{X}$, and that u and v are two nodes in the graph. Suppose further that $(v, u) \notin \mathcal{E}$ but that that edge exists in the Delaunay graph of $\mathcal{X}$.

If we could sample a query point q such that $\delta(q, u) < \delta(q, v)$ but $\delta(q, w) > \max(\delta(q, u), \delta(q, v))$ for all $w \neq u, v$, then we are done. That is because, if we entered the graph through v, then v is a local optimum in its neighborhood: all other points that are connected to v have a distance larger than $\delta(q, v)$. But v is not the globally optimal solution, so that the greedy traversal does not converge to the optimal solution.

It remains to show that such a point q always exists. Suppose it did not. That is, for any point that is in the Voronoi region of u, there is a data point $w \neq v$ that is closer to it than v. If that were the case, then no ball whose boundary passes through u and v can be empty, which contradicts Lemma 6.1 (the "empty-circle" property of the Delaunay graph). □

As a final remark on the Delaunay graph and its use in top-1 retrieval, we note that the Delaunay graph only makes sense if we have precise knowledge of the structure of the space (i.e., the metric). It is not enough to have just pairwise distances between points in a collection $\mathcal{X}$. In fact, Navarro [2002] showed that if pairwise distances are all we know about a collection of points, then the only sensible graph that contains the Delaunay graph and is amenable to greedy search is the complete graph. This is stated as the following theorem.

Theorem 6.3 *Suppose the structure of the metric space is unknown, but we have pairwise distances between the points in a collection* $\mathcal{X}$*, due to an arbitrary, but proper distance function* δ*. For every choice of* $u, v \in \mathcal{X}$*, there is a choice of the metric space such that* $(u, v) \in \mathcal{E}$*, where* $G = (\mathcal{V}, \mathcal{E})$ *is a Delaunay graph for* $\mathcal{X}$*.*

Proof. The idea behind the proof is to assume $(u, v) \notin \mathcal{E}$, then construct a query point that necessitates the existence of an edge between u and v. To that end, consider a query point q such that its distance to u is $C + \epsilon$ for some constant C and $\epsilon > 0$, its distance to v is C, and its distance to every other point in $\mathcal{X}$ is $C + 2\epsilon$.

This is a valid arrangement if we choose ϵ such that $\epsilon \leq 1/2 \min_{x,y \in \mathcal{X}} \delta(x, y)$ and C such that $C \geq 1/2 \max_{x,y \in \mathcal{X}} \delta(x, y)$. It is easy to verify that, if those conditions hold, a point q with the prescribed distances can exist as the distances do not violate any of the triangle inequalities.

Consider then a search starting from node u. If $(u, v) \notin \mathcal{E}$, then for the search algorithm to walk from u to the optimal solution, v, it must first get *farther* from q. But we know by the properties of the Delaunay graph that such an event implies that u (which would be the local optimum) must be the global optimum. That is clearly not true. So we must have that $(u, v) \in \mathcal{E}$, giving the claim. □

6.2.4 Top-k Retrieval

Let us now consider the general case of top-k retrieval over the Delaunay graph. The following result states that Algorithm 3 is correct if executed on any graph that contains the Delaunay graph, in the sense that it returns the optimal solution to top-k retrieval.

Theorem 6.4 *Let $G = (\mathcal{V}, \mathcal{E})$ be a graph that contains the Delaunay graph of m vectors $\mathcal{X} \subset \mathbb{R}^d$. Algorithm 3 over G gives the optimal solution to the top-k retrieval problem for any arbitrary query q if $\delta(\cdot, \cdot)$ is proper.*

Proof. As with the proof of Theorem 6.1, we show the result for the case where δ is the Euclidean distance and leave the proof of the more general case as an exercise.

The proof is similar to the proof of Theorem 6.1 but the argument needs a little more care when $k > 1$. Suppose Algorithm 3 for q stops at some local optimum set Q that is different from the global optimum, Q^*. In other words, $Q \bigtriangleup Q^* \neq \emptyset$ where $\bigtriangleup$ denotes the symmetric difference between sets.

Let $r = \max_{u \in Q} \delta(q, u)$ and consider the ball $B(q, r)$. Because $Q \bigtriangleup Q^* \neq \emptyset$, there must be at least k points in the interior of this ball. Let $v \notin Q$ be a point in the interior and suppose $u \in Q$ is its closest point in the ball. Clearly, the ball $B((u + v)/2, \delta(u, v)/2)$ is empty: otherwise v would not be the closest point to u. By Lemma 6.1, we must have that $(u, v) \in \mathcal{E}$. This is a contradiction because Algorithm 3 would, before termination, place v in Q to replace the node that is on the surface of the ball. □

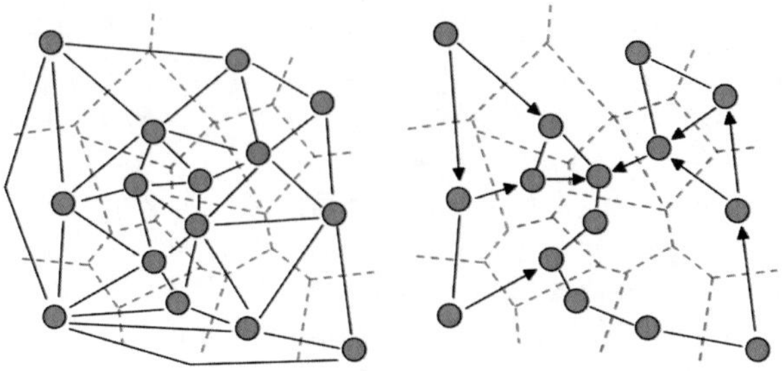

(a) Delaunay graph (b) 2-NN graph

Fig. 6.4: Comparison of the Delaunay graph (a) with the k-NN graph for $k = 2$ (b) for an example collection in $\mathbb{R}^2$. In the illustration of the directed k-NN graph, edges that go in both directions are rendered as lines without arrow heads. Notice that, the top left node cannot be reached from the rest of the graph.

6.2.5 The k-NN Graph

From our discussion of Voronoi diagrams and Delaunay graphs, it appears as though we have found the graph we have been looking for. Indeed, the Delaunay graph of a collection of vectors gives us the exact solution to top-k queries, using such a strikingly simple search algorithm. Sadly, the story does not end there and, as usual, the relentless curse of dimensionality poses a serious challenge.

The first major obstacle in high dimensions actually concerns the construction of the Delaunay graph itself. While there are many algorithms [Edelsbrunner and Shah, 1992, Guibas et al., 1992, Guibas and Stolfi, 1985] that can be used to construct the Delaunay graph—or, to be more precise, to perform Delaunay *triangulation*—all suffer from an exponential dependence on the number of dimensions d. So building the graph itself seems infeasible when d is too large.

Even if we were able to quickly construct the Delaunay graph for a large collection of points, we would face a second debilitating issue: The graph is close to complete! While exact bounds on the expected number of edges in the graph surely depend on the data distribution, in high dimensions the graph becomes necessarily more dense. Consider, for example, vectors that are independent and identically-distributed in each dimension. Recall from our discussion from Chapter 2, that in such an arrangement of points, the distance between any pair of points tends to concentrate sharply. As a result, the Delaunay graph has an edge between almost every pair of nodes.

These two problems are rather serious, rendering the guarantees of the Delaunay graph for top-k retrieval mainly of theoretical interest. These same difficulties motivated research to *approximate* the Delaunay graph. One prominent method is known as the k-NN graph [Chávez and Tellez, 2010, Hajebi et al., 2011, Fu and Cai, 2016].

The k-NN graph is simply a k-regular graph where every node (i.e., vector) is connected to its k closest nodes. So $(u, v) \in \mathcal{E}$ if $v \in \arg\min^{(k)}_{w \in \mathcal{X}} \delta(u, w)$. Note that, the resulting graph may be directed, depending on the choice of δ. We should mention, however, that researchers have explored ways of turning the k-NN graph into an undirected graph [Chávez and Tellez, 2010]. An example is depicted in Figure 6.4.

We must remark on two important properties of the k-NN graph. First, the graph itself is far more efficient to construct than the Delaunay graph [Chen et al., 2009, Vaidya, 1989, Connor and Kumar, 2010, Dong et al., 2011]. The second point concerns the connectivity of the graph. As Brito et al. [1997] show, under mild conditions governing the distribution of the vectors and with k large enough, the resulting graph has a high probability of being connected. When k is too small, on the other hand, the resulting graph may become too sparse, leading the greedy search algorithm to get stuck in local minima.

Finally, at the risk of stating the obvious, the k-NN graph does not enjoy any of the guarantees of the Delaunay graph in the context of top-k retrieval. That is simply because the k-NN graph is likely only a sub-graph of the Delaunay graph, while Theorems 6.1 and 6.4 are provable only for super-graphs of the Delaunay graph. Despite these deficiencies, the k-NN graph remains an important component of advanced graph-based, approximate top-k retrieval algorithms.

6.2.6 The Case of Inner Product

Everything we have stated so far about the Voronoi diagrams and its duality with the Delaunay graph was contingent on $\delta(\cdot, \cdot)$ being proper. In particular, the proof of the optimality guarantees implicitly require non-negativity and the triangle inequality. As a result, none of the results apply to MIPS *prima facie*. As it turns out, however, we could extend the definition of Voronoi regions and the Delaunay graph to inner product, and present guarantees for MIPS (with $k = 1$, but not with $k > 1$). That is the proposal by Morozov and Babenko [2018].

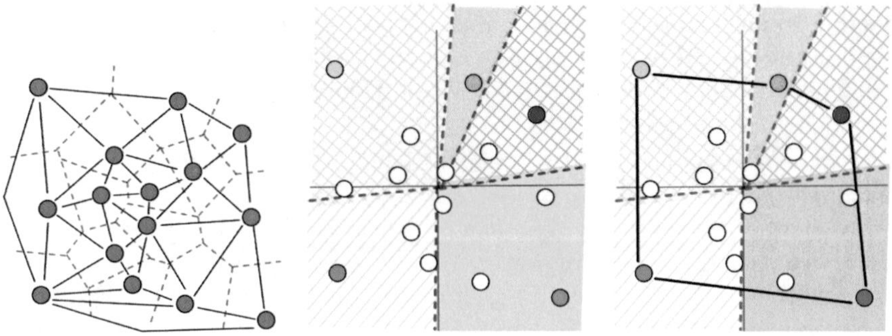

(a) Euclidean Delaunay (b) Inner Product Voronoi (c) IP-Delaunay

Fig. 6.5: Comparison of the Voronoi diagrams and Delaunay graphs for the same set of points but according to Euclidean distance versus inner product. Note that, for the non-metric distance function based on inner product, the Voronoi regions are convex cones determined by the intersection of half-spaces passing through the origin. Observe additionally that the inner product-induced Voronoi region of a point (those in white) may be an empty set. Such points can never be the solution to the 1-MIPS problem.

6.2.6.1 The IP-Delaunay Graph

Let us begin by characterizing the Voronoi regions for inner product. The Voronoi region $\mathcal{R}_u$ of a vector $u \in \mathcal{X}$ comprises of the set of points for which u is the maximizer of inner product:

$$\mathcal{R}_u = \{x \in \mathbb{R}^d \mid u = \arg\max_{v \in \mathcal{X}} \langle x, v \rangle\}.$$

This definition is essentially the same as how we defined the Voronoi region for a proper δ, and, indeed, the resulting Voronoi diagram is a partitioning of the whole space. The properties of the resulting Voronoi regions, however, could not be more different.

First, recall from Section 1.3.3 that inner product does not even enjoy what we called coincidence. That is, in general, $u = \arg\max_{v \in \mathcal{X}} \langle u, v \rangle$ is not guaranteed. So it is very much possible that $\mathcal{R}_u$ is empty for some $u \in \mathcal{X}$. Second, when $\mathcal{R}_u \neq \emptyset$, it is a convex cone that is the intersection of half-spaces that *pass through the origin*. So the Voronoi regions have a substantially different geometry. Figure 6.5(b) visualizes this phenomenon.

Moving on to the Delaunay graph, Morozov and Babenko [2018] construct the graph in much the same way as before and call the resulting graph the IP-Delaunay graph. Two nodes $u, v \in \mathcal{V}$ in the IP-Delaunay graph are connected if their Voronoi regions intersect: $\mathcal{R}_u \cap \mathcal{R}_v \neq \emptyset$. Note that, by the reasoning above, the nodes whose Voronoi regions are empty will be isolated in the graph. These nodes represent vectors that can never be the solution to MIPS for any query—remember that we are only considering $k = 1$. So it would be inconsequential if we removed these nodes from the graph. This is also illustrated in Figure 6.5(c).

Considering the data structure above for inner product, Morozov and Babenko [2018] prove the following result to give optimality guarantee for the greedy search algorithm for 1-MIPS (granted we enter the graph from a non-isolated node). Nothing, however, may be said about k-MIPS.

Theorem 6.5 *Suppose $G = (\mathcal{V}, \mathcal{E})$ is a graph that contains the IP-Delaunay graph for a collection $\mathcal{X}$ minus the isolated nodes. Invoking Algorithm 3 with $k = 1$ and $\delta(\cdot, \cdot) = -\langle \cdot, \cdot \rangle$ gives the optimal solution to the top-1 MIPS problem.*

Proof. If we showed that a local optimum is necessarily the global optimum, then we are done. To that end, consider a query q for which Algorithm 3 terminates when it reaches node $u \in \mathcal{X}$ which is distinct from the globally optimal solution $u^* \notin N(u)$. In other words, we have that $\langle q, u \rangle > \langle q, v \rangle$ for all $v \in N(u)$, but $\langle q, u^* \rangle > \langle q, u \rangle$ and $(u, u^*) \notin \mathcal{E}$. If that is true, then $q \notin \mathcal{R}_u$, the Voronoi region of u, but instead we must have that $q \in \mathcal{R}_{u^*}$.

Now define the collection $\tilde{\mathcal{X}} \triangleq N(u) \cup \{u\}$, and consider the Voronoi diagram of the resulting collection. It is easy to show that the Voronoi region of u in the presence of points in $\tilde{\mathcal{X}}$ is the same as its region given $\mathcal{X}$. From before, we also know that $q \notin \mathcal{R}_u$. Considering the fact that $\mathbb{R}^d = \bigcup_{v \in \tilde{\mathcal{X}}} \mathcal{R}_v$, q must belong to $\mathcal{R}_v$ for some $v \in \tilde{\mathcal{X}}$ with $v \neq u$. That implies that $\langle q, v \rangle > \langle q, u \rangle$ for some $v \in \tilde{\mathcal{X}} \setminus u$. But because $v \in N(u)$ (by construction), the last inequality poses a contradiction to our premise that u was locally optimal. □

In addition to the fact that the IP-Delaunay graph does not answer top-k queries, it also suffers from the same deficiencies we noted for the Euclidean Delaunay graph earlier in this section. Naturally then, to make the data structure more practical in high-dimensional regimes, we must resort to heuristics and approximations, which in their simplest form may be the k-MIPS graph (i.e., a k-NN graph where the distance function for finding the top-k nodes is inner product). This is the general direction Morozov and Babenko [2018] and a few other works have explored [Liu et al., 2019, Zhou et al., 2019].

As in the case of metric distance functions, none of the guarantees stated above port over to these approximate graphs. But, once again, empirical evidence gathered from a variety of datasets show that these graphs perform reasonably well in practice, even for top-k with $k > 1$.

6.2.6.2 Is the IP-Delaunay Graph Necessary?

Morozov and Babenko [2018] justify the need for developing the IP-Delaunay graph by comparing its structure with the following alternative: First, apply a MIPS-to-NN *asymmetric* transformation [Bachrach et al., 2014] from $\mathbb{R}^d$ to $\mathbb{R}^{d+1}$. This involves transforming a data point u with $\phi_d(u) = [u;\ \sqrt{1 - \|u\|_2^2}]$ and a query point q with $\phi_q(v) = [v; 0]$. Next, construct the standard (Euclidean) Delaunay graph over the transformed vectors.

What happens if we form the Delaunay graph on the transformed collection $\phi_d(\mathcal{X})$? Observe the Euclidean distance between $\phi_d(u)$ and $\phi_d(v)$ for two vectors $u, v \in \mathcal{X}$:

$$\begin{aligned}\|\phi_d(u) - \phi_d(v)\|_2^2 &= \|\phi_d(u)\|_2^2 + \|\phi_d(v)\|_2^2 - 2\langle \phi_d(u), \phi_d(v)\rangle \\ &= \left(\|u\|_2^2 + 1 - \|u\|_2^2\right) + \left(\|v\|_2^2 + 1 - \|v\|_2^2\right) \\ &\quad - 2\langle u, v\rangle - 2\sqrt{\left(1 - \|u\|_2^2\right)\left(1 - \|v\|_2^2\right)}.\end{aligned}$$

Should we use these distances to construct the Delaunay graph, the resulting structure will have nothing to do with the original MIPS problem. That is because the L_2 distance between a pair of transformed data points is not rank-equivalent to the inner product between the original data points. For this reason, Morozov and Babenko [2018] argue that the IP-Delaunay graph is a more sensible choice.

However, we note that their argument rests heavily on their particular choice of MIPS-to-NN transformation. The transformation they chose makes sense in contexts where we *only* care about preserving the inner product between query-data point pairs. But when forming the Delaunay graph, preserving inner product between pairs of data points, too, is imperative. That is the reason why we lose rank-equivalence between L_2 in $\mathbb{R}^{d+1}$ and inner product in $\mathbb{R}^d$.

There are, in fact, MIPS-to-NN transformations that are more appropriate for this problem and would invalidate the argument for the need for the IP-Delaunay graph. Consider for example $\phi_d : \mathbb{R}^d \rightarrow \mathbb{R}^{d+m}$ for a collection $\mathcal{X}$ of m vectors as follows: $\phi_d(u^{(i)}) = u^{(i)} \oplus \left(\sqrt{1 - \|u^{(i)}\|_2^2}\right) e_{(d+i)}$, where $u^{(i)}$ is the i-th data point in the collection, and e_j is the j-th standard basis vector. In other words, the i-th d-dimensional data point is augmented with an m-dimensional sparse vector whose i-th coordinate is non-zero. The query transformation is simply $\phi_q(q) = q \oplus \mathbf{0}$, where $\mathbf{0} \in \mathbb{R}^m$ is a vector of m zeros.

Despite the dependence on m, the transformation is remarkably easy to manage: the sparse subspace of every vector has at most one non-zero coordinate, making the doubling dimension of the sparse subspace $\mathcal{O}(\log m)$ by Lemma 3.5. Distance computation between the transformed vectors, too, has negligible overhead. Crucially, we regain rank-equivalence between L_2 distance in $\mathbb{R}^{d+m}$ and inner product in $\mathbb{R}^d$ not only for query-data point pairs, but also for pairs of data points:

$$\begin{aligned}\|\phi_d(u) - \phi_d(v)\|_2^2 &= \|\phi_d(u)\|_2^2 + \|\phi_d(v)\|_2^2 - 2\langle \phi_d(u), \phi_d(v)\rangle \\ &= 2 - 2\langle u, v\rangle.\end{aligned}$$

Finally, unlike the IP-Delaunay graph, the standard Delaunay graph in $\mathbb{R}^{d+m}$ over the transformed vector collection has optimality guarantee for the top-k retrieval problem per Theorem 6.4. It is, as such, unclear if the IP-Delaunay graph is even necessary as a theoretical tool.

In other words, suppose we are given a collection of points $\mathcal{X}$ and inner product as the similarity function. Consider a graph index where the presence of an edge is decided based on the inner product between data points. Take another graph index built for the transformed $\mathcal{X}$ using the transformation described above from $\mathbb{R}^d$ to $\mathbb{R}^{d+m}$, and where the edge set is formed on the basis of the Euclidean distance between two (transformed) data points. The two graphs are equivalent.

The larger point is that, MIPS over m points in $\mathbb{R}^d$ is equivalent to NN over a transformation of the points in $\mathbb{R}^{d+m}$. While the transformation increases the apparent dimensionality, the intrinsic dimensionality of the data only increases by $\mathcal{O}(\log m)$.

6.3 The Small World Phenomenon

Consider, once again, the Delaunay graph but, for the moment, set aside the fact that it is a prohibitively-expensive data structure to maintain for high dimensional vectors. By construction, every node in the graph is only connected to its Voronoi neighbors (i.e., nodes whose Voronoi region intersects with the current node's). We showed that such a topology affords *navigability*, in the sense that the greedy procedure in Algorithm 3 can traverse the graph only based on information about immediate neighbors of a node and yet arrive at the globally optimal solution to the top-k retrieval problem.

Let us take a closer look at the traversal algorithm for the case of $k = 1$. It is clear that, navigating from the entry node to the solution takes us through every Voronoi region along the path. That is, we cannot "skip" a Voronoi region that lies between the entry node and the answer. This implies that the running time of Algorithm 3 is directly affected by the diameter of the graph (in addition to the average degree of nodes).

Can we enhance this topology by adding *long-range* edges between non-Voronoi neighbors, so that we may skip over a fraction of Voronoi regions? After all, Theorem 6.4 guarantees navigability so long as the graph *contains* the Delaunay graph. Starting with the Delaunay graph and inserting long-range edges, then, will not take away any of the guarantees. But, what is the

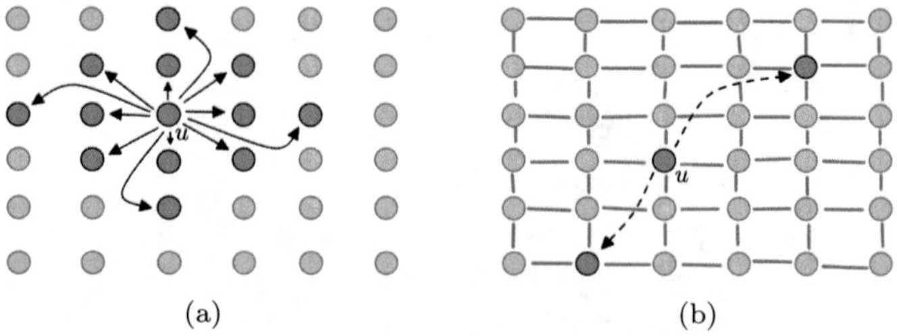

Fig. 6.6: Example graphs generated by the probabilistic model introduced by Kleinberg [2000]. (a) illustrates the directed edges from u for the following configuration: $r = 2$, $l = 0$. (b) renders the regular structure for $r = 1$, where edges without arrows are bi-directional, and the long-range edges for node u with configuration $l = 2$.

right number of long-range edges and how do we determine which remote nodes should be connected? This section reviews the theoretical results that help answer these questions.

6.3.1 Lattice Networks

Let us begin with a simple topology that is relatively easy to reason about—we will see later how the results from this section can be generalized to the Delaunay graph. The graph we have in mind is a lattice network where $m \times m$ nodes are laid on a two-dimensional grid. Define the distance between two nodes as their *lattice (Manhattan) distance* (i.e., the minimal number of horizontal and vertical hops that connect two nodes). That is the network examined by Kleinberg [2000] in a seminal paper that studied the effect of long-range edges on the time complexity of Algorithm 3.

We should take a brief detour and note that, Kleinberg [2000], in fact, studied the problem of *transmitting* a message from a source to a known target using the best-first-search algorithm, and quantified the average number of hops required to do that in the presence of a variety of classes of long-range edges. That, in turn, was inspired by a social phenomenon colloquially known as the "small-world phenomenon": The empirical observation that two strangers are linked by a short chain of acquaintances [Milgram, 1967, Jeffrey Travers, 1969].

In particular, Kleinberg [2000] was interested in explaining why and under what types of long-range edges should our greedy algorithm be able to navigate to the optimal solution, by only utilizing information about immediate neighbors. To investigate this question, Kleinberg [2000] introduced the following probabilistic model of the lattice topology as an abstraction of individuals and their social connections.

6.3.1.1 The Probabilistic Model

Every node in the graph has a (directed) edge with every other node within lattice distance r, for some fixed hyperparameter $r \geq 1$. These connections make up the regular structure of the graph. Overlaid with this structure is a set of random, long-range edges that are generated according to the following probabilistic model. For fixed constants $l \geq 0$ and $\alpha \geq 0$, we insert a directed edge between every node u and l other nodes, where a node $v \neq u$ is selected with probability proportional to $\delta(u, v)^{-\alpha}$ where $\delta(u, v) = \|u - v\|_1$ is the lattice distance. Example graphs generated by this process are depicted in Figure 6.6.

The model above is reasonably powerful as it can express a variety of topologies. For example, when $l = 0$, the resulting graph has no long-range edges. When $l > 0$ and $\alpha = 0$, then every node $v \neq u$ in the graph has an equal chance of being the destination of a long-range edge from u. When α is large, the protocol becomes more biased to forming a long-range connection from u to nodes closer to it.

6.3.1.2 The Claim

Kleinberg [2000] shows that, when $0 \leq \alpha < 2$, the best-first-search algorithm must visit at least $\mathcal{O}_{r,l,\alpha}\big(m^{(2-\alpha)/3}\big)$ nodes. When $\alpha > 2$, the number of nodes visited is at least $\mathcal{O}_{r,l,\alpha}\big(m^{(\alpha-2)/(\alpha-1)}\big)$ instead. But, rather uniquely, when $\alpha = 2$ and $r = l = 1$, we visit a number of nodes that is at most poly-logarithmic in m.

Theorem 6.6 states this result formally. But before we present the theorem, we state a useful lemma.

Lemma 6.2 *Generate a lattice* $G = (\mathcal{V}, \mathcal{E})$ *of* $m \times m$ *nodes using the probabilistic model above with* $\alpha = 2$ *and* $l = 1$. *The probability that there exists a long-range edge between two nodes* $u, v \in \mathcal{V}$ *is at least* $\delta(u, v)^{-2}/4\ln(6m)$.

Proof. u chooses $v \neq u$ as its long-range destination with the following probability: $\delta(u, v)^{-2}/\sum_{w \neq u} \delta(u, w)^{-2}$. Let us first bound the denominator as follows:

$$\sum_{w \neq u} \delta(u,w)^{-2} \leq \sum_{i=1}^{2m-2} (4i)(i^{-2}) = 4 \sum_{i=1}^{2m-2} \frac{1}{j}$$
$$\leq 4 + 4\ln\left(2m-2\right) \leq 4\ln\left(6m\right).$$

In the expression above, we derived the first inequality by iterating over all possible (lattice) distances between m^2 nodes on a two-dimensional grid (ranging from 1 to $2m-2$ if u and w are at diagonally opposite corners), and noticing that there are at most $4i$ nodes at distance i from node u. From this we infer that the probability that node $(u,v) \in \mathcal{E}$ is at least $\delta(u,v)^{-2}/4\ln(6m)$. □

Theorem 6.6 *Generate a lattice $G = (\mathcal{V}, \mathcal{E})$ of $m \times m$ nodes using the probabilistic model above with $\alpha = 2$ and $r = l = 1$. The best-first-search algorithm beginning from any arbitrary node and ending in a target node visits at most $\mathcal{O}(\log^2 m)$ nodes on average.*

Proof. Define a sequence of sets A_i, where each A_i consists of nodes whose distance to the target u^* is greater than 2^i and at most 2^{i+1}. Formally, $A_i = \{v \in \mathcal{V} \mid 2^i < \delta(u^*,v) \leq 2^{i+1}\}$. Suppose that the algorithm is currently in node u and that $\log m \leq \delta(u,u^*) < m$, so that $u \in A_i$ for some $\log\log m \leq i < \log m$. What is the probability that the algorithm exits the set A_i in the next step?

That happens when one of u's neighbors has a distance to u^* that is at most 2^i. In other words, u must have a neighbor that is in the set $A_{<i} = \cup_{j=0}^{j=i-1} A_j$. The number of nodes in $A_{<i}$ is at least:

$$1 + \sum_{s=1}^{2^i} s = 1 + \frac{2^i\left(2^i+1\right)}{2} > 2^{2i-1}.$$

How likely is it that $(u,v) \in \mathcal{E}$ if $v \in A_{<i}$? We apply Lemma 6.2, noting that the distance of each of the nodes in $A_{<i}$ with u is at most $2^{i+1}+2^i < 2^{i+2}$. We obtain that, the probability that u is connected to a node in $A_{<i}$ is at least $2^{2i-1}(2^{i+2})^{-2}/4\ln(6m) = 1/128\ln(6m)$.

Next, consider the total number of nodes in A_i that are visited by the algorithm, and denote it by X_i. In expectation, we have the following:

$$\mathbb{E}\left[X_i\right] = \sum_{j=1}^{\infty} \mathbb{P}\left[X_i \geq j\right] \leq \sum_{j=1}^{\infty} \left(1 - \frac{1}{128\ln(6m)}\right)^{j-1} = 128\ln(6m).$$

We obtain the same bound if we repeat the arguments for $i = \log m$. When $0 \leq i < \log\log m$, the algorithm visits at most $\log m$ nodes, so that the bound above is trivially true.

Denoting by X the total number of nodes visited, $X = \sum_{j=0}^{\log m} X_j$, we conclude that:

$$\mathbb{E}\left[X\right] \leq (1 + \log m)(128 \ln(6m)) = \mathcal{O}(\log^2 m),$$

thereby completing the proof. □

The argument made by Kleinberg [2000] is that, in a lattice network where each node is connected to its (at most four) nearest neighbors within unit distance, and where every node has a long-range edge to one other node with probability that is proportional to $1/\delta(\cdot,\cdot)^2$, then the greedy algorithm visits at most a poly-logarithmic number of nodes. Translating this result to the case of top-1 retrieval using Algorithm 3 over the same network, we can state that the time complexity of the algorithm is $\mathcal{O}(\log^2 m)$, because the total number of neighbors per node is $\mathcal{O}(1)$.

While this result is significant, it only holds for the lattice network with the lattice distance. It has thus no bearing on the time complexity of top-1 retrieval over the Delaunay graph with the Euclidean distance. In the next section, we will see how Beaumont et al. [2007a] close this gap.

6.3.2 Extension to the Delaunay Graph

We saw in the preceding section that, the secret to creating a provably navigable graph where the best-first-search algorithm visits a poly-logarithmic number of nodes in the lattice network, was the highly specific distribution from which long-range edges were sampled. That element turns out to be the key ingredient when extending the results to the Delaunay graph too, as Beaumont et al. [2007a] argue.

We will describe the algorithm for data in the two-dimensional unit square. That is, we assume that the collection of data points $\mathcal{X}$ and query points are in $[0,1]^2$. That the vectors are bounded is not a limitation *per se*—as we discussed previously, we can always normalize vectors into the hypercube without loss of generality. That the algorithm does not naturally extend to high dimensions is a serious limitation, but then again, that is not surprising considering the Delaunay graph is expensive to construct. However, in the next section, we will review heuristics that take the idea to higher dimensions.

6.3.2.1 The Probabilistic Model

Much like the lattice network, we assume there is a base graph and a number of randomly generated long-range edges between nodes. For the base graph, Beaumont et al. [2007a] take the Delaunay graph.[1] As for the long-range edges, each node has a directed edge to one other node that is selected

[1] Beaumont et al. [2007a] additionally connect all nodes that are within $\delta_{\text{MIN}} \propto 1/m$ distance from each other, where δ_{MIN} is chosen such that the expected number of uniformly-

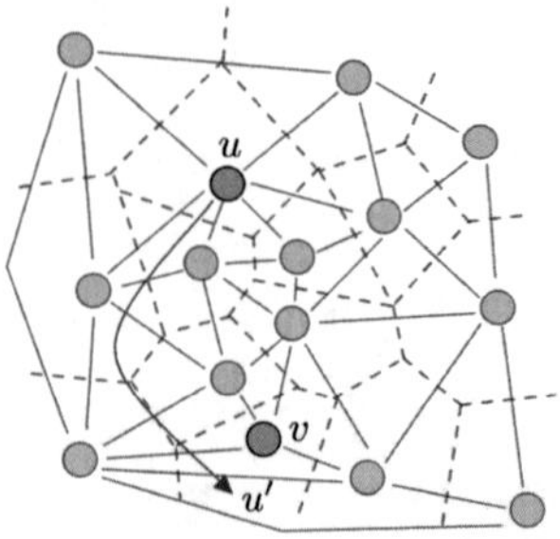

Fig. 6.7: Formation of a long-range edge from u to v by the probabilistic model of Beaumont et al. [2007a]. First, we jump from u to a random long-range end-point u', then designate its nearest neighbor (v) as the target of the edge.

at random, following a process we will describe shortly. Observe that in this model, the number of neighbors of each node is $\mathcal{O}(1)$.

We already know from Theorem 6.4 that, because the network above contains the Delaunay graph, it is navigable by Algorithm 3. What remains to be investigated is what type of long-range edges could reduce the number of hops (i.e., the number of nodes the algorithm must visit as it navigates from an entry node to a target node). Because at each hop the algorithm needs to evaluate distances with $\mathcal{O}(1)$ neighbors, improving the number of steps directly improves the time complexity of Algorithm 3 (for the case of $k = 1$).

Beaumont et al. [2007a] show that, if long-range edges are chosen according to the following protocol, then the number of hops is poly-logarithmic in m. The protocol is simple: For a node u in the graph, first sample α uniformly from $[\ln \delta^*, \ln \delta_*]$, where $\delta^* = \min_{v,w \in \mathcal{X}} \delta(v, w)$ and $\delta_* = \max_{v,w \in \mathcal{X}} \delta(v, w)$. Then choose $\theta \sim [0, 2\pi]$, to finally obtain $u' = u + z$ where z is the vector $[e^\alpha \cos\theta, e^\alpha \sin\theta]$. Let us refer to u' as the "long-range end-point," and note that this point may escape the $[0, 1]^2$ square. Next, we find the nearest node v to u' and add a directed edge from u to v. This is demonstrated in Figure 6.7.

6.3.2.2 The Claim

Given the resulting graph, Beaumont et al. [2007a] state and prove that the average number of hops taken by the best-first-search algorithm is poly-logarithmic. Before we discuss the claim, however, let us state a useful lemma.

Lemma 6.3 *The probability that the long-range end-point from a node u lands in a ball centered at another node v with radius $\beta\delta(u, v)$ for some small*

distributed points in a ball of radius δ_{MIN} is 1. We reformulate their method without δ_{MIN} in the present monograph to simplify their result.

$\beta \in [0,1]$ is at least $K\beta^2/(1+\beta)^2$ where $K = (2\ln \Delta)^{-1}$ and $\Delta = \delta_*/\delta^*$ is the aspect ratio.

Proof. The probability that the long-range end-point lands in an area dS that covers the distance $[r, r+dr]$ and angle $[\theta, \theta + d\theta]$, for small dr and $d\theta$, is:

$$\frac{d\theta}{2\pi}\frac{\ln(r+dr) - \ln r}{\ln \delta_* - \ln \delta^*} \approx \frac{d\theta}{2\pi}\frac{dr/r}{\ln \Delta} = \frac{1}{2\pi \ln \Delta}\frac{rd\theta dr}{r^2} \approx \frac{dS}{2\pi \ln \Delta r^2}$$

.

Observe now that the distance between a point u and any point in the ball described in the lemma is at most $(1+\beta)\delta(u,v)$. We can therefore see that the probability that the long-range end-point lands in $B(v, \beta\delta(u,v))$ is at least:

$$\frac{\pi\beta^2\delta(u,v)^2}{2\pi \ln \Delta(1+\beta)^2\delta(u,v)^2} = \frac{\beta^2}{2\ln(\Delta)(1+\beta)^2},$$

as required. □

Theorem 6.7 *Generate a graph $G = (\mathcal{V}, \mathcal{E})$ according to the probabilistic model described above, for vectors in $[0,1]^2$ equipped with the Euclidean distance $\delta(\cdot,\cdot)$. The number of nodes visited by the best-first-search algorithm starting from any arbitrary node and ending at a target node is $\mathcal{O}(\log^2 \Delta)$.*

Proof. The proof follows the same reasoning as in the proof of Theorem 6.6.

Suppose we are currently at node u and that u^* is our target node. By Lemma 6.3, the probability that the long-range end-point of u lands in $B(u^*, \delta(u,u^*)/6)$ is at least $1/98\ln \Delta$. As such, the total number of hops, X, from u to a point in $B(u^*, \delta(u,u^*)/6)$ has the following expectation:

$$\mathbb{E}[X] = \sum_{i=1}^{\infty} \mathbb{P}[X \geq i] \leq \sum_{i=1}^{\infty} \left(1 - \frac{1}{98\ln \Delta}\right)^{i-1} = 98\ln \Delta.$$

Every time the algorithm moves from the current node u to some other node in $B(u^*, \delta(u,u^*)/6)$, the distance is shrunk by a factor of $6/5$. As such, the total number of hops in expectation is at most:

$$\left(\ln_{6/5} \Delta\right) \times \left(98\ln \Delta\right) = \mathcal{O}(\log^2 \Delta).$$

□

We highlight that, Beaumont et al. [2007a] choose the interval from which α is sampled differently. Indeed, α in their work is chosen uniformly from the range $\delta_{\text{MIN}} \propto 1/m$ and $\sqrt{2}$. Substituting that configuration into Theorem 6.7 gives an expected number of hops that is $\mathcal{O}(\log^2 m)$.

6.3.3 Approximation

The results of Beaumont et al. [2007a] are encouraging. In theory, so long as we can construct the Delaunay graph, we not only have the optimality guarantee, but we are also guaranteed to have a poly-logarithmic number of hops to reach the optimal answer. Alas, as we have discussed previously, the Delaunay graph is expensive to build in high dimensions.

Moreover, the number of neighbors per node is no longer $\mathcal{O}(1)$. So even if we inserted long-range edges into the Delaunay graph, it is not immediate if the time saved by skipping Voronoi regions due to long-range edges offsets the additional time the algorithm spends computing distances between each node along the path and its neighbors.

We are back, then, to approximation with the help of heuristics. Beaumont et al. [2007b] describe one such method in a follow-up study. Their method approximates the Voronoi regions of every node by resorting to a *gossip* protocol. In this procedure, every node has a list of $3d + 1$ of its current neighbors, where d denotes the dimension of the space. In every iteration of the algorithm, every node passes its current list to its neighbors. When a node receives this information, it takes the union of all lists, and finds the subset of $3d + 1$ points with the minimal volume. This subset becomes the node's current list of neighbors. While a naïve implementation of the protocol is prohibitively expensive, Beaumont et al. [2007b] discuss an alternative to estimating the volume induced by a set of $3d + 1$ points, and the search for the minimal volume.

Malkov et al. [2014] take a different approach. They simply permute the vectors in the collection $\mathcal{X}$, and sequentially add each vector to the graph. Every time a vector is inserted into the graph, it is linked to its k nearest neighbors from the current snapshot of the graph. The intuition is that, as the graph grows, the edges added earlier in the evolution of the graph serve as long-range edges in the final graph, and the more recent edges form an approximation of the k-NN graph, which itself is an approximation of the Delaunay graph. Later Malkov and Yashunin [2020] modify the algorithm by introducing a hierarchy of graphs. The resulting graph has proved successful in practice and, despite its lack of theoretical guarantees, is both effective and highly efficient.

6.4 Neighborhood Graphs

In the preceding section, our starting point was the Delaunay graph. We augmented it with random long-range connections to improve the transmission rate through the network. Because the resulting structure contains the Delaunay graph, we get the optimality guarantee of Theorem 6.4 for free. But, as a result of the complexity of the Delaunay construction in high dimensions, we had to approximate the structure instead, losing all guarantees in the process. Frustratingly, the approximate structure obtained by the heuristics we discussed, is certainly not a super-graph of the Delaunay graph, nor is it necessarily its sub-graph. In fact, even the fundamental property of connectedness is not immediately guaranteed. There is therefore nothing meaningful to say about the theoretical behavior of such graphs.

In this section, we do the opposite. Instead of adding edges to the Delaunay graph and then resorting to heuristics to create a completely different graph, we prune the edges of the Delaunay graph to find a structure that is its *sub-graph.* Indeed, we cannot say anything meaningful about the optimality of *exact* top-k retrieval, but as we will later see, we can state formal results for the approximate top-k retrieval variant—albeit in a very specific case. The structure we have in mind is known as the Relative Neighborhood Graph (RNG) [Toussaint, 1980, Jaromczyk and Toussaint, 1992].

In an RNG, $G = (\mathcal{V}, \mathcal{E})$, for a distance function $\delta(\cdot, \cdot)$, there is an undirected edge between two nodes $u, v \in \mathcal{V}$ if and only if $\delta(u, v) < \max\big(\delta(u, w), \delta(w, v)\big)$ for all $w \in \mathcal{V} \setminus \{u, v\}$. That is, the graph guarantees that, if $(u, v) \in \mathcal{E}$, then there is no other point in the collection that is simultaneously closer to u and v, than u and v are to each other. Conceptually, then, we can view constructing an RNG as pruning away edges in the Delaunay graph that violate the RNG property.

The RNG was shown to contain the Minimum Spanning Tree [Toussaint, 1980], so that it is guaranteed to be connected. It is also provably contained in the Delaunay graph [O'Rourke, 1982] in any metric space and in any number of dimensions. As a final property, it is not hard to see that such a graph G comes with a weak optimality guarantee for the best-first-search algorithm: If $q = u^* \in \mathcal{X}$, then the greedy traversal algorithm returns the node associated with q, no matter where it enters the graph. That is due simply to the following fact: If the current node u is a local optimum but not the global optimum, then there must be an edge connecting u to a node that is closer to u^*. Otherwise, u itself must be connected to u^*.

Later, Arya and Mount [1993] proposed a *directed* variant of the RNG, which they call the *Sparse Neighborhood Graph* (SNG) that is arguably more suitable for top-k retrieval. For every node $u \in \mathcal{V}$, we apply the following procedure: Let $\mathcal{U} = \mathcal{V} \setminus \{u\}$. Sort the nodes in $\mathcal{U}$ in increasing distance to u. Then, remove the closest node (say, v) from $\mathcal{U}$ and add an edge between u to v. Finally, remove from $\mathcal{U}$ all nodes w that satisfy $\delta(u, w) > \delta(w, v)$. The process is repeated until $\mathcal{U}$ is empty. It can be immediately seen that the weak optimality guarantee from before still holds in the SNG.

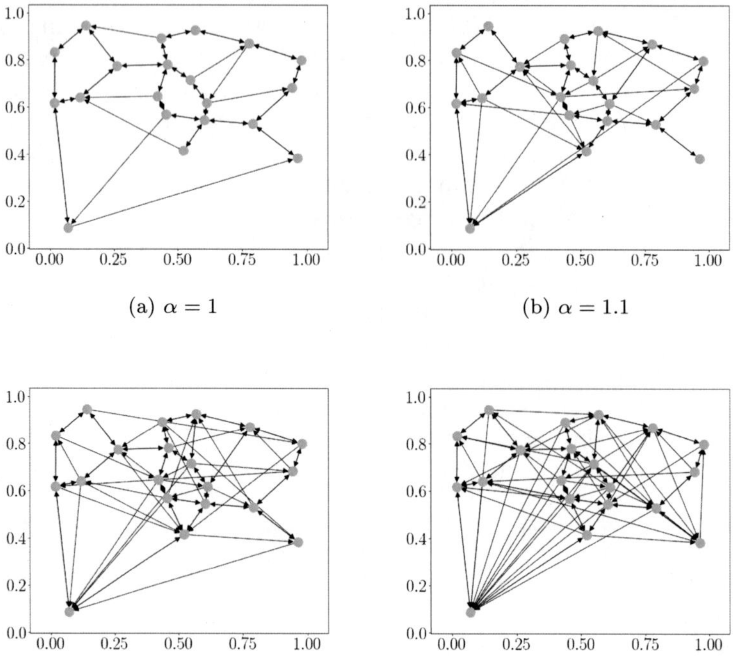

(a) $\alpha = 1$

(b) $\alpha = 1.1$

(c) $\alpha = 1.2$

(d) $\alpha = 1.3$

Fig. 6.8: Examples of α-SNGs on a dataset of 20 points drawn uniformly from $[0, 1]^2$ (blue circles). When $\alpha = 1$, we recover the standard SNG. As α becomes larger, the resulting graph becomes more dense.

Neighborhood graphs are the backbone of many graph algorithms for top-k retrieval [Malkov et al., 2014, Malkov and Yashunin, 2020, Harwood and Drummond, 2016, Fu et al., 2019, 2022, Jayaram Subramanya et al., 2019]. While many of these algorithms make for efficient methods in practice, the Vamana construction [Jayaram Subramanya et al., 2019] stands out as it introduces a novel super-graph of the SNG that turns out to have provable theoretical properties. That super-graph is what Indyk and Xu [2023] call an *α-shortcut reachable* SNG, which we will review next. For brevity, though, we call this graph simply α-SNG.

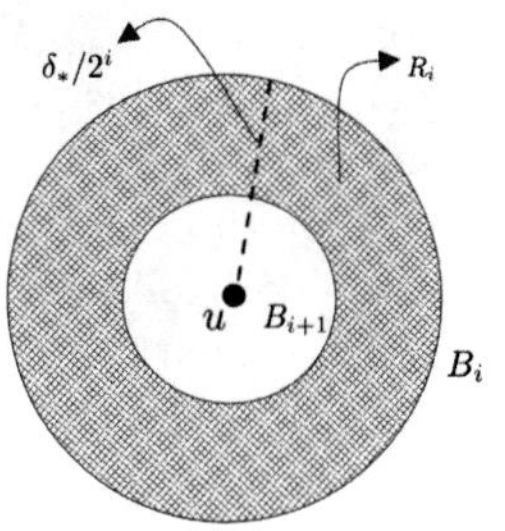

Fig. 6.9: The sets B_i and rings R_i in the proof of Theorem 6.8.

6.4.1 From SNG to α-SNG

Jayaram Subramanya et al. [2019] introduce a subtle adjustment to the SNG construction. In particular, suppose we are processing a node u, have already extracted the node v whose distance to u is minimal among the nodes in $\mathcal{U}$ (i.e., $v = \arg\min_{w \in \mathcal{U}} \delta(u, w)$), and are now deciding which nodes to discard from $\mathcal{U}$. In the standard SNG construction, we remove a node w for which $\delta(u, w) > \delta(w, v)$. But in the modified construction, we instead discard a node w if $\delta(u, w) > \alpha\delta(w, v)$ for some $\alpha > 1$. Note that, the case of $\alpha = 1$ simply gives the standard SNG. Figure 6.8 shows a few examples of α-SNGs on a toy dataset.

That is what Indyk and Xu [2023] later refer to as an α-shortcut reachable graph. They define α-shortcut reachability as the property where, for any node u, we have that every other node w is either the target of an edge from u (so that $(u, w) \in \mathcal{E}$), or that there is a node v such that $(u, v) \in \mathcal{E}$ and $\delta(u, w) \geq \alpha\delta(w, v)$. Clearly, the graph constructed by the procedure above is α-shortcut reachable by definition.

6.4.1.1 Analysis

Indyk and Xu [2023] present an analysis of the α-SNG for a collection of vectors $\mathcal{X}$ with *doubling dimension* $d_\circ$ as defined in Definition 3.2.

For collections with a fixed doubling constant, Indyk and Xu [2023] state two bounds. One gives a bound on the degree of every node in an α-SNG. The other tells us the expected number of hops from any arbitrary entry node to an ϵ-approximate solution to top-1 queries. The two bounds together give us an idea of the time complexity of Algorithm 3 over an α-SNG as well as its accuracy.

Theorem 6.8 *The degree of any node in an α-SNG is $\mathcal{O}\big((4\alpha)^{d_\circ} \log \Delta\big)$ if the collection $\mathcal{X}$ has doubling dimension $d_\circ$ and aspect ratio $\Delta = \delta_*/\delta^*$.*

Proof. Consider a node $u \in \mathcal{V}$. For each $i \in [\log_2 \Delta]$, define a ball centered at u with radius $\delta_*/2^i$: $B_i = B(u, \delta_*/2^i)$. From this, construct rings $R_i = B_i \backslash B_{i+1}$. See Figure 6.9 for an illustration.

Because $\mathcal{X}$ has a constant doubling dimension, we can cover each R_i with $\mathcal{O}\big((4\alpha)^{d_\circ}\big)$ balls of radius $\delta_*/\alpha 2^{i+2}$. By construction, two points in each of these cover balls are at most $\delta_*/\alpha 2^{i+1}$ apart. At the same time, the distance from u to any point in a cover ball is at least $\delta_*/2^{i+1}$. By construction, all points in a cover ball except one are discarded as we form u's edges in the α-SNG. As such, the total number of edges from u is bounded by the total number of cover balls, which is $\mathcal{O}\big((4\alpha)^{d_\circ} \log \Delta\big)$. □

Theorem 6.9 *If $G = (\mathcal{V}, \mathcal{E})$ is an α-SNG for collection $\mathcal{X}$, then Algorithm 3 with $k = 1$ returns an $(\frac{\alpha+1}{\alpha-1} + \epsilon)$-approximate top-1 solution by visiting $\mathcal{O}\big(\log_\alpha \frac{\Delta}{(\alpha-1)\epsilon}\big)$ nodes.*

Proof. Suppose q is a query point and $u^* = \arg\min_{u \in \mathcal{X}} \delta(q, u)$. Further assume that the best-first-search algorithm is currently in node v_i with distance $\delta(q, v_i)$ to the query. We make the following observations:

- By triangle inequality, we know that $\delta(v_i, u^*) \leq \delta(v_i, q) + \delta(q, u^*)$; and,
- By construction of the α-SNG, v_i is either connected to u^* or to another node whose distance to u^* is shorter than $\delta(v_i, u^*)/\alpha$.

We can conclude that, the distance from q to the next node the algorithm visits, v_{i+1}, is at most:

$$\begin{aligned} \delta(v_{i+1}, q) &\leq \delta(v_{i+1}, u^*) + \delta(u^*, q) \\ &\leq \frac{\delta(v_i, u^*)}{\alpha} + \delta(u^*, q) \\ &\leq \frac{\delta(v_i, q)}{\alpha} + (\alpha + 1)\delta(q, u^*). \end{aligned}$$

By induction, we see that, if the entry node is $s \in \mathcal{V}$:

$$\begin{aligned} \delta(v_i, q) &\leq \frac{\delta(s, q)}{\alpha^i} + (\alpha + 1)\delta(q, u^*) \sum_{j=1}^{i} \alpha^{-j} \\ &\leq \frac{\delta(s, q)}{\alpha^i} + \frac{\alpha + 1}{\alpha - 1}\delta(q, u^*). \end{aligned} \tag{6.1}$$

There are three cases to consider.

Case 1: When $\delta(s, q) > 2\delta_*$, then by triangle inequality, $\delta(q, u^*) > \delta(s, q) - \delta(s, u^*) > \delta(s, q) - \delta_* > \delta(s, q)/2$. Plugging this into Equation (6.1) yields:

$$\delta(v_i, q) \le \frac{2\delta(q, u^*)}{\alpha^i} + \frac{\alpha+1}{\alpha-1}\delta(q, u^*)$$
$$\implies \frac{\delta(v_i, q)}{\delta(q, u^*)} \le \frac{2}{\alpha^i} + \frac{\alpha+1}{\alpha-1}.$$

As such, for any $\epsilon > 0$, the algorithm returns a $\big(\frac{\alpha+1}{\alpha-1}+\epsilon\big)$-approximate solution in $\log_\alpha 2/\epsilon$ steps.

Case 2: $\delta(s, q) \le 2\delta_*$ and $\delta(q, u^*) \ge \frac{\alpha-1}{4(\alpha+1)}\delta^*$. By Equation (6.1), the algorithm returns a $\big(\frac{\alpha+1}{\alpha-1} + \epsilon\big)$-approximate solution as soon as $\delta(s, q)/\alpha^i < \epsilon\delta(q, u^*)$. So in this case:

$$\frac{\delta(v_i, q)}{\delta(q, u^*)} \le \frac{2\delta_*}{\alpha^i\delta(q, u^*)} + \frac{\alpha+1}{\alpha-1}$$
$$\le \frac{8(\alpha+1)\delta_*}{\alpha^i(\alpha-1)\delta^*} + \frac{\alpha+1}{\alpha-1}.$$

As such, the number of steps to reach the approximation level is $\log_\alpha \frac{8(\alpha+1)\Delta}{(\alpha-1)\epsilon}$ which is $\mathcal{O}\big(\log_\alpha \Delta/(\alpha-1)\epsilon\big)$.

Case 3: $\delta(s, q) \le 2\delta_*$ and $\delta(q, u^*) < \frac{\alpha-1}{4(\alpha+1)}\delta^*$. Suppose $v_i \ne u^*$. Observe that: (a) $\delta(v_i, u^*) \ge \delta^*$; (b) $\delta(v_i, q) > \delta(q, u^*)$; and (c) $\delta(q, u^*) < \delta^*/2$ by assumption. As such, triangle inequality gives us: $\delta(v_i, q) > \delta(v_i, u^*) - \delta(u^*, q) > \delta^* - \delta^*/2 = \delta^*/2$. Together with Equation (6.1), we obtain:

$$\frac{\delta^*}{2} \le \delta(v_i, q) \le \frac{2\delta_*}{\alpha^i} + \frac{\delta^*}{4}$$
$$\implies \alpha^i \le 8\Delta \implies i \le \log_\alpha 8\Delta.$$

The three cases together give the desired result. □

In addition to the bounds above, Indyk and Xu [2023] present negative results for other major SNG-based graph algorithms by proving (via contrived examples) linear-time lower-bounds on their performance. These results together show the significance of the pruning parameter α in the α-SNG construction.

6.4.1.2 Practical Construction of α-SNGs

The algorithm described earlier to construct an α-SNG for m points has $\mathcal{O}(m^3)$ time complexity. That is too expensive for even moderately large values of m. That prompted Jayaram Subramanya et al. [2019] to approximate the α-SNG by way of heuristics.

The starting point in the approximate construction is a random R-regular graph: Every node is connected to R other nodes selected at random. The algorithm then processes each node in random order as follows. Given node u, it begins by searching the current snapshot of the graph for the top L nodes for the query point u, using Algorithm 3. Denote the returned set of nodes by $\mathcal{S}$. It then performs the pruning algorithm by setting $\mathcal{U} = \mathcal{S} \setminus \{u\}$, rather than $\mathcal{U} = \mathcal{V} \setminus \{u\}$. That is the gist of the modified construction procedure.[2]

Naturally, we lose all guarantees for approximate top-k retrieval as a result [Indyk and Xu, 2023]. We do, however, obtain a more practical algorithm instead that, as the authors show, is both efficient and effective.

6.5 Closing Remarks

This chapter deviated from the pattern we got accustomed to so far in the monograph. The gap between theory and practice in Chapters 4 and 5 was narrow or none. That gap is rather wide, on the other hand, in graph-based retrieval algorithms. Making theory work in practice required a great deal of heuristics and approximations.

Another major departure is the activity in the respective bodies of literature. Whereas trees and hash families have reached a certain level of maturity, the literature on graph algorithms is still evolving, actively so. A quick search through scholarly articles shows growing interest in this class of algorithms. This monograph itself presented results that were obtained very recently.

There is good reason for the uptick in research activity. Graph algorithms are among the most successful algorithms there are for top-k vector retrieval. They are often remarkably fast during retrieval and produce accurate solution sets.

That success makes it all the more enticing to improve their other characteristics. For example, graph indices are often large, requiring far too much memory. Incorporating compression into graphs, therefore, is a low-hanging fruit that has been explored [Singh et al., 2021] but needs further investigation. More importantly, finding an even sparser graph without losing accuracy is key in reducing the size of the graph to begin with, and that boils down to designing better heuristics.

Heuristics play a key role in the construction time of graph indices too. Building a graph index for a collection of billions of points, for example, is not feasible for the variant of the Vamana algorithm that offers theoretical guarantees. Heuristics introduced in that work lost all such guarantees, but made the graph more practical.

Enhancing the capabilities of graph indices too is an important practical consideration. For example, when the graph is too large and, so, must rest on disk, optimizing disk access is essential in maintaining the speed of query processing [Jayaram Subramanya et al., 2019]. When the collection of vectors

---

[2] We have omitted minor but important details of the procedure in our prose. We refer the interested reader to [Jayaram Subramanya et al., 2019] for a description of the full algorithm.

is live and dynamic, the graph index must naturally handle deletions and insertions in real-time [Singh et al., 2021]. When vectors come with metadata and top-k retrieval must be constrained to the vectors that pass a certain set of metadata filters, then a greedy traversal of the graph may prove suboptimal [Gollapudi et al., 2023]. All such questions warrant extensive (often applied) research and go some way to make graph algorithms more attractive to production systems.

There is thus no shortage of practical research questions. However, the aforementioned gap between theory and practice should not dissuade us from developing better theoretical algorithms. The models that explained the small world phenomenon may not be directly applicable to top-k retrieval in high dimensions, but they inspired heuristics that led to the state of the art. Finding theoretically-sound edge sets that improve over the guarantees offered by Vamana could form the basis for other, more successful heuristics too.

References

S. Arya and D. M. Mount. Approximate nearest neighbor queries in fixed dimensions. In *Proceedings of the 4th Annual ACM-SIAM Symposium on Discrete Algorithms*, pages 271–280, 1993.

Y. Bachrach, Y. Finkelstein, R. Gilad-Bachrach, L. Katzir, N. Koenigstein, N. Nice, and U. Paquet. Speeding up the xbox recommender system using a euclidean transformation for inner-product spaces. In *Proceedings of the 8th ACM Conference on Recommender Systems*, page 257–264, 2014.

O. Beaumont, A.-M. Kermarrec, L. Marchal, and E. Riviere. Voronet: A scalable object network based on voronoi tessellations. In *2007 IEEE International Parallel and Distributed Processing Symposium*, pages 1–10, 2007a.

O. Beaumont, A.-M. Kermarrec, and É. Rivière. Peer to peer multidimensional overlays: Approximating complex structures. In *Principles of Distributed Systems*, pages 315–328, 2007b.

M. Brito, E. Chávez, A. Quiroz, and J. Yukich. Connectivity of the mutual k-nearest-neighbor graph in clustering and outlier detection. *Statistics & Probability Letters*, 35(1):33–42, 1997.

E. Chávez and E. S. Tellez. Navigating k-nearest neighbor graphs to solve nearest neighbor searches. In *Proceedings of the 2nd Mexican Conference on Pattern Recognition: Advances in Pattern Recognition*, pages 270–280, 2010.

J. Chen, H.-r. Fang, and Y. Saad. Fast approximate knn graph construction for high dimensional data via recursive lanczos bisection. *Journal of Machine Learning Research*, 10:1989–2012, 12 2009.

M. Connor and P. Kumar. Fast construction of k-nearest neighbor graphs for point clouds. *IEEE Transactions on Visualization and Computer Graphics*, 16(4):599–608, 2010.

B. Delaunay. Sur la sphère vide. *Bulletin de l'Académie des Sciences de l'URSS. Classe des sciences mathématiques et na*, 1934(6):793–800, 1934.

W. Dong, C. Moses, and K. Li. Efficient k-nearest neighbor graph construction for generic similarity measures. In *Proceedings of the 20th International Conference on World Wide Web*, pages 577–586, 2011.

H. Edelsbrunner and N. R. Shah. Incremental topological flipping works for regular triangulations. In *Proceedings of the 8th Annual Symposium on Computational Geometry*, pages 43–52, 1992.

S. Fortune. *Voronoi Diagrams and Delaunay Triangulations*, pages 377–388. CRC Press, Inc., 1997.

C. Fu and D. Cai. Efanna : An extremely fast approximate nearest neighbor search algorithm based on knn graph, 2016.

C. Fu, C. Xiang, C. Wang, and D. Cai. Fast approximate nearest neighbor search with the navigating spreading-out graph. *Proceedings of the VLDB Endowment*, 12(5):461–474, 1 2019.

C. Fu, C. Wang, and D. Cai. High dimensional similarity search with satellite system graph: Efficiency, scalability, and unindexed query compatibility. *IEEE Transactions on Pattern Analysis and Machine Intelligence*, 44(8): 4139–4150, 2022.

S. Gollapudi, N. Karia, V. Sivashankar, R. Krishnaswamy, N. Begwani, S. Raz, Y. Lin, Y. Zhang, N. Mahapatro, P. Srinivasan, A. Singh, and H. V. Simhadri. Filtered-diskann: Graph algorithms for approximate nearest neighbor search with filters. In *Proceedings of the ACM Web Conference 2023*, pages 3406–3416, 2023.

L. Guibas and J. Stolfi. Primitives for the manipulation of general subdivisions and the computation of voronoi. *ACM Transactions on Graphics*, 4 (2):74–123, 04 1985.

L. J. Guibas, D. E. Knuth, and M. Sharir. Randomized incremental construction of delaunay and voronoi diagrams. *Algorithmica*, 7(1–6):381–413, 3 1992.

K. Hajebi, Y. Abbasi-Yadkori, H. Shahbazi, and H. Zhang. Fast approximate nearest-neighbor search with k-nearest neighbor graph. In *Twenty-Second International Joint Conference on Artificial Intelligence*, 2011.

B. Harwood and T. Drummond. Fanng: Fast approximate nearest neighbour graphs. In *2016 IEEE Conference on Computer Vision and Pattern Recognition*, pages 5713–5722, 2016.

P. Indyk and H. Xu. Worst-case performance of popular approximate nearest neighbor search implementations: Guarantees and limitations. In *Proceedings of the 36th Conference on Neural Information Processing Systems*, 2023.

J. Jaromczyk and G. Toussaint. Relative neighborhood graphs and their relatives. *Proceedings of the IEEE*, 80(9):1502–1517, 1992.

S. Jayaram Subramanya, F. Devvrit, H. V. Simhadri, R. Krishnawamy, and R. Kadekodi. Diskann: Fast accurate billion-point nearest neighbor search on a single node. In *Advances in Neural Information Processing Systems*, volume 32, 2019.

S. M. Jeffrey Travers. An experimental study of the small world problem. *Sociometry*, 32(4):425–443, 1969.

J. Kleinberg. The small-world phenomenon: An algorithmic perspective. In *Proceedings of the 32nd Annual ACM Symposium on Theory of Computing*, pages 163–170, 2000.

W. Li, Y. Zhang, Y. Sun, W. Wang, M. Li, W. Zhang, and X. Lin. Approximate nearest neighbor search on high dimensional data — experiments, analyses, and improvement. *IEEE Transactions on Knowledge and Data Engineering*, 32(8):1475–1488, 2020.

J. Liu, X. Yan, X. DAI, Z. Li, J. Cheng, and M. Yang. Understanding and improving proximity graph based maximum inner product search. In *AAAI Conference on Artificial Intelligence*, 2019.

Y. Malkov, A. Ponomarenko, A. Logvinov, and V. Krylov. Approximate nearest neighbor algorithm based on navigable small world graphs. *Information Systems*, 45:61–68, 2014.

Y. A. Malkov and D. A. Yashunin. Efficient and robust approximate nearest neighbor search using hierarchical navigable small world graphs. *IEEE Transactions on Pattern Analysis and Machine Intelligence*, 42(4):824–836, 4 2020.

S. Milgram. The Small-World Problem. *Psychology Today*, 1(1):61–67, 1967.

S. Morozov and A. Babenko. Non-metric similarity graphs for maximum inner product search. In *Proceedings of the 32nd International Conference on Neural Information Processing Systems*, pages 4726–4735, 2018.

G. Navarro. Searching in metric spaces by spatial approximation. *The VLDB Journal*, 11(1):28–46, 08 2002.

J. O'Rourke. Computing the relative neighborhood graph in the l1 and l∞ metrics. *Pattern Recognition*, 15(3):189–192, 1982.

A. Singh, S. J. Subramanya, R. Krishnaswamy, and H. V. Simhadri. Freshdiskann: A fast and accurate graph-based ann index for streaming similarity search, 2021.

G. T. Toussaint. The relative neighbourhood graph of a finite planar set. *Pattern Recognition*, 12(4):261–268, 1980.

P. M. Vaidya. Ano(n logn) algorithm for the all-nearest-neighbors problem. *Discrete and Computational Geometry*, 4(2):101–115, 12 1989.

M. Wang, X. Xu, Q. Yue, and Y. Wang. A comprehensive survey and experimental comparison of graph-based approximate nearest neighbor search. *Proceedings of the VLDB Endowment*, 14(11):1964–1978, jul 2021.

Z. Zhou, S. Tan, Z. Xu, and P. Li. Möbius transformation for fast inner product search on graph. In *Advances in Neural Information Processing Systems*, volume 32, 2019.

Chapter 7
Clustering

Abstract We have seen index structures that manifest as trees, hash tables, and graphs. In this chapter, we will introduce a fourth way of organizing data points: clusters. It is perhaps the most natural and the simplest of the four methods, but also the least theoretically-justified. We will see why that is as we describe the details of clustering-based algorithms to top-k retrieval.

7.1 Algorithm

As usual, we begin by indexing a collection of m data points $\mathcal{X} \subset \mathbb{R}^d$. Except in this paradigm, that involves invoking a **clustering** function, $\zeta : \mathbb{R}^d \rightarrow [C]$, that is appropriate for the distance function $\delta(\cdot, \cdot)$, to map every data point to one of C clusters, where C is an arbitrary parameter. A typical choice for ζ is the KMeans algorithm with $C = \mathcal{O}(\sqrt{m})$. We then organize $\mathcal{X}$ into a *table* whose row i records the subset of points that are mapped to the i-th cluster: $\zeta^{-1}(i) \triangleq \{u \mid u \in \mathcal{X},\ \zeta(u) = i\}$.

Accompanying the index is a **routing** function $\tau : \mathbb{R}^d \rightarrow [C]^\ell$. It takes an arbitrary point q as input and returns ℓ clusters that are more likely to contain the nearest neighbor of q with respect to δ. In a typical instance of this framework $\tau(\cdot)$ is defined as follows:

$$\tau(q) = \underset{i \in [C]}{\overset{(\ell)}{\arg\min}}\, \delta\Bigg(q,\ \underbrace{\frac{1}{|\zeta^{-1}(i)|} \sum_{u \in \zeta^{-1}(i)} u}_{\mu_i}\Bigg), \tag{7.1}$$

where μ_i is the *centroid* of the i-th cluster. In other words, $\tau(\cdot)$ simply solves the top-ℓ retrieval problem over the collection of centroids! We will assume that τ is defined as above in the remainder of this section.

S. Bruch, *Foundations of Vector Retrieval*, https://doi.org/10.1007/978-3-031-55182-6_7

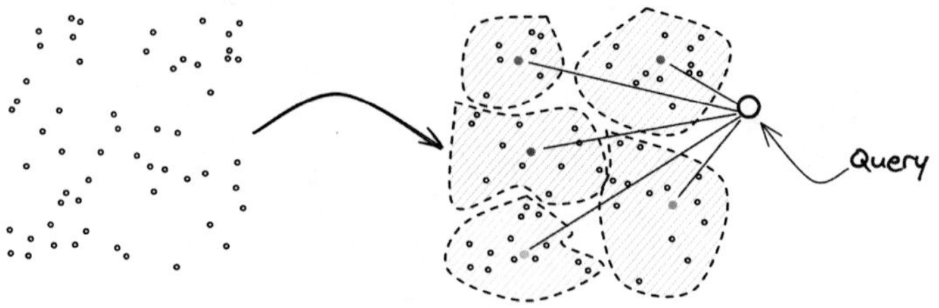

Fig. 7.1: Illustration of the clustering-based retrieval method. The collection of points (left) is first partitioned into clusters (regions enclosed by dashed boundary on the right). When processing a query q using Equation (7.1), we compute $\delta(q, \cdot)$ for the centroid (solid circles) of every cluster and conduct our search over the ℓ "closest" clusters.

When processing a query q, we take a two-step approach. We first obtain the list of clusters returned by $\tau(q)$, then solve the top-k retrieval problem over the union of the identified clusters. Figure 7.1 visualizes this procedure.

Notice that, the search for top-ℓ clusters by using Equation (7.1) and the secondary search over the clusters identified by τ are themselves instances of the approximate top-k retrieval problem. The parameter C determines the amount of effort that must be spent in each of the two phases of search: When $C = 1$, the cluster retrieval problem is solved trivially, whereas as $C \to \infty$, cluster retrieval becomes equivalent to top-k retrieval over the entire collection. Interestingly, these operations can be delegated to a subroutine that itself uses a tree-, hash-, graph-, or even a clustering-based solution. That is, a clustering-based approach can be easily paired with any of the previously discussed methods!

This simple protocol—with some variant of KMeans as ζ and τ as in Equation (7.1)—works well in practice [Auvolat et al., 2015, Jégou et al., 2011, Bruch et al., 2023b, Babenko and Lempitsky, 2012, Chierichetti et al., 2007]. We present the results of our own experiments on various real-world datasets in Figure 7.2. This method owes its success to the empirical phenomenon that real-world data points tend to follow a multi-modal distribution, naturally forming clusters around each mode. By identifying these clusters and grouping data points together, we reduce the search space at the expense of retrieval quality.

However, to date, no formal analysis has been presented to quantify the retrieval error. The choice of ζ and τ, too, have been left largely unexplored, with KMeans and Equation (7.1) as default answers. It is, for example, not known if KMeans is the right choice for a given δ. Or, whether clustering with spillage, where each data point may belong to multiple clusters, might reduce the overall error, as it did in Spill Trees. It is also an open question if, for a particular choice of ζ and δ, there exists a more effective routing

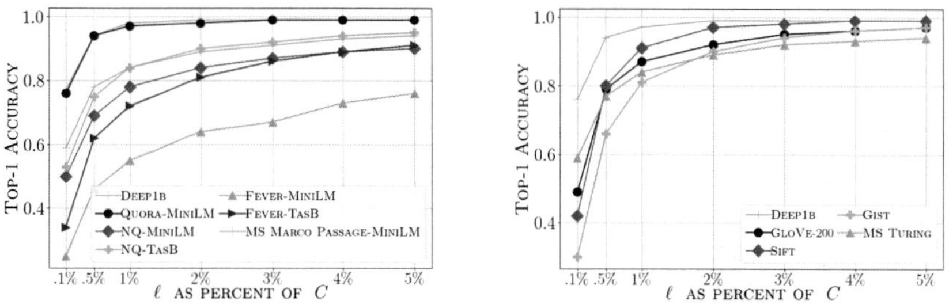

(a) MIPS (b) NN

Fig. 7.2: Performance of the clustering-based retrieval method on various real-world collections, described in Appendix A. The figure shows top-1 accuracy versus the number of clusters, ℓ, considered by the routing function $\tau(\cdot)$ as a percentage of the number of clusters C. In these experiments, we set $C = \sqrt{m}$, where $m = |\mathcal{X}|$ is the size of the collection, and use spherical KMeans (MIPS) and standard KMeans (NN) to form clusters.

function—including learnt functions tailored to a query distribution—that uses higher-order statistics from the cluster distributions.

In spite of these shortcomings, the algorithmic framework above contains a fascinating insight that is actually useful for a rather different end-goal: vector compression, or more precisely, *quantization*. We will unpack this connection in Chapter 9.

7.2 Closing Remarks

This chapter departed entirely from the theme of this monograph. Whereas we are generally able to say something intelligent about trees, hash functions, and graphs, top-k retrieval by clustering has emerged entirely based on our intuition that data points naturally form clusters. We cannot formally determine, for example, the behavior of the retrieval system as a function of the clustering algorithm itself, the number of clusters, or the routing function. All that must be determined empirically.

What we do observe often in practice, however, is that clustering-based top-k retrieval is *efficient* [Paulevé et al., 2010, Auvolat et al., 2015, Bruch et al., 2023b, Jégou et al., 2011], at least in the case of Nearest Neighbor search with Euclidean distance, where KMeans is a theoretically appropriate choice. It is efficient in the sense that retrieval accuracy often reaches an acceptable level after probing a few top-ranking clusters as identified by Equation (7.1).

That we have a method that is efficient in practice, but its efficiency and the conditions under which it is efficient are unexplained, constitutes a substantial gap and thus presents multiple consequential open questions. These questions involve optimal clustering, routing, and bounds on retrieval accuracy.

When the distance function is the Euclidean distance and our objective is to learn the Voronoi regions of data points, the KMeans clustering objective makes sense. We can even state formal results regarding the optimality of the resulting clustering [Arthur and Vassilvitskii, 2007]. That argument is no longer valid when the distance function is based on inner product, where we must learn the inner product Voronoi cones, and where some points may have an empty Voronoi region. What objective we must optimize for MIPS, therefore, is an open question that, as we saw in this chapter, has been partially explored in the past [Guo et al., 2020].

Even when we know what the right clustering algorithm is, there is still the issue of "balance" that we must understand how to handle. It would, for example, be far from ideal if the clusters end up having very different sizes. Unfortunately, that happens quite naturally if the data points have highly variable norms and the clustering algorithm is based on KMeans: Data points with large norms become isolated, while vectors with small norms form massive clusters.

What has been left entirely untouched is the routing machinery. Equation (7.1) is the *de facto* routing function, but one that is possibly suboptimal. That is because, Equation (7.1) uses the mean of the data points within a cluster as the representative or *sketch* of that cluster. When clusters are highly concentrated around their mean, such a sketch accurately reflects the potential of each cluster. But when clusters have different shapes, higher-order statistics from the cluster may be required to accurately route queries to clusters.

So the question we are faced with is the following: What is a good sketch of each cluster? Is there a *coreset* of data points within each cluster that lead to better routing of queries during retrieval? Can we quantify the probability of error—in the sense that the cluster containing the optimal solution is not returned by the routing function—given a sketch?

We may answer these questions differently if we had some idea of what the query distribution looks like. Assuming access to a set of training queries, it may be possible to learn a more optimal sketch using supervised learning methods. Concepts from learning-to-rank [Bruch et al., 2023a] seem particularly relevant to this setup. To see how, note that the outcome of the routing function is to identify the cluster that contains the optimal data point for a query. We could view this as ranking clusters with respect to a query, where we wish for the "correct" cluster to appear at the top of the ranked list. Given this mental model, we can evaluate the quality of a routing function using any of the many ranking quality metrics such as Reciprocal Rank (defined as the reciprocal of the rank of the correct cluster). Learning a ranking function that maximizes Reciprocal Rank can then be done indirectly by optimizing a custom cross entropy-based surrogate, as proved by Bruch et al. [2019] and Bruch [2021].

Perhaps the more important open question is understanding when clustering is efficient and why. Answering that question may require exploring the connection between clustering-based top-k retrieval, branch-and-bound algorithms, and LSH.

Take any clustering algorithm, ζ. If one could show formally that ζ behaves like an LSH family, then clustering-based top-k retrieval simply collapses to LSH. In that case, not only do the results from that literature apply, but the techniques developed for LSH (such as multi-probe LSH) too port over to clustering.

Similarly, one may adopt the view that finding the top cluster is a series of decisions, each determining which side of a hyperplane a point falls. Whereas in Random Partition Trees or Spill Trees, such decision hyperplanes were random directions, here the hyperplanes are correlated. Nonetheless, that insight could help us produce clusters with spillage, where data points belong to multiple clusters, and in a manner that helps reduce the overall error.

References

D. Arthur and S. Vassilvitskii. K-means++: The advantages of careful seeding. In *Proceedings of the Eighteenth Annual ACM-SIAM Symposium on Discrete Algorithms*, pages 1027–1035, 2007.

A. Auvolat, S. Chandar, P. Vincent, H. Larochelle, and Y. Bengio. Clustering is efficient for approximate maximum inner product search, 2015.

A. Babenko and V. Lempitsky. The inverted multi-index. In *2012 IEEE Conference on Computer Vision and Pattern Recognition*, pages 3069–3076, 2012.

S. Bruch. An alternative cross entropy loss for learning-to-rank. In *Proceedings of the Web Conference 2021*, page 118–126, 2021.

S. Bruch, X. Wang, M. Bendersky, and M. Najork. An analysis of the softmax cross entropy loss for learning-to-rank with binary relevance. In *Proceedings of the 2019 ACM SIGIR International Conference on Theory of Information Retrieval*, page 75–78, 2019.

S. Bruch, C. Lucchese, and F. M. Nardini. Efficient and effective tree-based and neural learning to rank. *Foundations and Trends in Information Retrieval*, 17(1):1–123, 2023a.

S. Bruch, F. M. Nardini, A. Ingber, and E. Liberty. Bridging dense and sparse maximum inner product search, 2023b.

F. Chierichetti, A. Panconesi, P. Raghavan, M. Sozio, A. Tiberi, and E. Upfal. Finding near neighbors through cluster pruning. In *Proceedings of the Twenty-Sixth ACM SIGMOD-SIGACT-SIGART Symposium on Principles of Database Systems*, pages 103–112, 2007.

R. Guo, P. Sun, E. Lindgren, Q. Geng, D. Simcha, F. Chern, and S. Kumar. Accelerating large-scale inference with anisotropic vector quantization. In *Proceedings of the 37th International Conference on Machine Learning*, 2020.

H. Jégou, M. Douze, and C. Schmid. Product quantization for nearest neighbor search. *IEEE Transactions on Pattern Analysis and Machine Intelligence*, 33(1):117–128, 2011.

L. Paulevé, H. Jégou, and L. Amsaleg. Locality sensitive hashing: A comparison of hash function types and querying mechanisms. *Pattern Recognition Letters*, 31(11):1348–1358, 2010.

Chapter 8
Sampling Algorithms

Abstract Nearly all of the data structures and algorithms we reviewed in the previous chapters are designed specifically for either nearest neighbor search or maximum cosine similarity search. MIPS is typically an afterthought. It is often cast as NN or MCS through a rank-preserving transformation and subsequently solved using one of these algorithms. That is so because inner product is not a proper metric, making MIPS different from the other vector retrieval variants. In this chapter, we review algorithms that are specifically designed for MIPS and that connect MIPS to the machinery underlying multi-arm bandits.

8.1 Intuition

That inner product is different can be a curse and a blessing. We have already discussed that curse at length, but in this chapter, we will finally learn something positive. And that is the fact that inner product is a linear function of data points and can be easily decomposed into its parts, thereby opening a unique path to solving MIPS.

The overarching idea in what we refer to as *sampling algorithms* is to avoid computing inner products. Instead, we either directly approximate the *likelihood* of a data point being the solution to MIPS (or, equivalently, its rank), or estimate its inner product with a query (i.e., its score). As we will see shortly, in both instances, we rely heavily on the linearity of inner product to estimate probabilities and derive bounds.

Approximating the ranks or scores of data points uses some form of sampling: we either sample data points according to a distribution defined by inner products, or sample a dimension to compute partial inner products with and eliminate sub-optimal data points iteratively. In the former, the more frequently a data point is sampled, the more likely it is to be the solution to MIPS. In the latter, the more dimensions we sample, the closer we

S. Bruch, *Foundations of Vector Retrieval*, https://doi.org/10.1007/978-3-031-55182-6_8

get to computing full inner products. Generally, then, the more samples we draw, the more accurate our solution to MIPS becomes.

An interesting property of using sampling to solve MIPS is that, regardless of *what* we are approximating, we can decide when to stop! That is, if we are given a time budget, we draw as many samples as our time budget allows and return our best guess of the solutions based on the information we have collected up to that point. The number of samples, in other words, serves as a knob that trades off accuracy for speed.

The remainder of this chapter describes these algorithms in much greater detail. Importantly, we will see how linearity makes the approximation-through-sampling feasible and efficient.

8.2 Approximating the Ranks

We are interested in finding the top-k data points with the largest inner product with a query $q \in \mathbb{R}^d$, from a collection $\mathcal{X} \subset \mathbb{R}^d$ of m points. Suppose that we had an efficient way of sampling a data point from $\mathcal{X}$ where the point $u \in \mathcal{X}$ has probability proportional to $\langle q, u \rangle$ of being selected.

If we drew a sufficiently large number of samples, the data point with the largest inner product with q would be selected most frequently. The data point with the second largest inner product would similarly be selected with the second highest frequency, and so on. So, if we counted the number of times each data point has been sampled, the resulting histogram would be a good approximation to the rank of each data point with respect to inner product with q.

That is the gist of the sampling algorithm we examine in this section. But while the idea is rather straightforward, making it work requires addressing a few critical gaps. The biggest challenge is drawing samples according to the distribution of inner products without actually computing any of the inner products! That is because, if we needed to compute $\langle q, u \rangle$ for all $u \in \mathcal{X}$, then we could simply sort data points accordingly and return the top-k; no need for sampling and the rest.

The key to tackling that challenge is the linearity of inner product. Following a few simple derivations using Bayes' theorem, we can break up the sampling procedure into two steps, each using marginal distributions only [Lorenzen and Pham, 2021, Ballard et al., 2015, Cohen and Lewis, 1997, Ding et al., 2019]. Importantly, one of these marginal distributions can be computed offline as part of indexing. That is the result we will review next.

8.2.1 Non-negative Data and Queries

We wish to draw a data point with probability that is proportional to its inner product with a query: $\mathbb{P}[u \mid q] \propto \langle q, u \rangle$. For now, we assume that $u, q \succeq 0$ for all $u \in \mathcal{X}$ and queries q.

Let us decompose this probability along each dimension as follows:

$$\mathbb{P}[u \mid q] = \sum_{t=1}^{d} \mathbb{P}[t \mid q]\, \mathbb{P}[u \mid t, q], \tag{8.1}$$

where the first term in the sum is the probability of sampling a dimension $t \in [d]$ and the second term is the likelihood of sampling u given a particular dimension. We can model each of these terms as follows:

$$\mathbb{P}[t \mid q] \propto \sum_{u \in \mathcal{X}} q_t u_t = q_t \sum_{u \in \mathcal{X}} u_t, \tag{8.2}$$

and,

$$\mathbb{P}[u \mid t, q] = \frac{\mathbb{P}[u \wedge t \mid q]}{\mathbb{P}[t \mid q]} \propto \frac{q_t u_t}{q_t \sum_{v \in \mathcal{X}} v_t} = \frac{u_t}{\sum_{v \in \mathcal{X}} v_t}. \tag{8.3}$$

In the above, we have assumed that $\sum_{v \in \mathcal{X}} v_t \neq 0$; if that sum is 0 we can simply discard the t-th dimension.

What we have done above allows us to draw a sample according to $\mathbb{P}[u \mid q]$ by, instead, drawing a dimension t according to $\mathbb{P}[t \mid q]$ first, then drawing a data point u according to $\mathbb{P}[u \mid t, q]$.

Sampling from these multinomial distribution requires constructing the distributions themselves. Luckily, $\mathbb{P}[u \mid t, q]$ is independent of q. Its distribution can therefore be computed offline: we create d tables, where the t-th table has m rows recording the probability of each data point being selected given dimension t using Equation (8.3). We can then use the alias method [Walker, 1977] to draw samples from these distributions using $\mathcal{O}(1)$ operations.

The distribution over dimensions given a query, $\mathbb{P}[t \mid q]$, must be computed online using Equation (8.2), which requires $\mathcal{O}(d)$ operations, assuming we compute $\sum_{u \in \mathcal{X}} u_t$ offline for each t and store them in our index. Again, using the alias method, we can subsequently draw samples with $\mathcal{O}(1)$ operations.

The procedure described above provides us with an efficient mechanism to perform the desired sampling. If we were to draw S samples, that could be done in $\mathcal{O}(d + S)$, where $\mathcal{O}(d)$ term is needed to construct the multinomial distribution that defines $\mathbb{P}[t \mid q]$.

As we draw samples, we maintain a histogram over the m data points, counting the number of times each point has been sampled. In the end, we can identify the top-k' (for $k' \geq k$) points based on these counts, compute their inner products with the query, and return the top-k points as the final solution set. All these operations together have time complexity $\mathcal{O}(d + S + m \log k' + k'd)$, with S typically being the dominant term.

8.2.2 The General Case

When the data points or queries may be negative, the algorithm described in the previous section will not work as is. To extend the sampling framework to general, real vectors, we must make a few minor adjustments.

First, we must ensure that the marginal distributions are valid. That is easy to do: In Equations (8.2) and (8.3), we replace each term with its absolute value. So, $\mathbb{P}[t \mid q]$ becomes proportional to $\sum_{u \in \mathcal{X}} |q_t u_t|$, and $\mathbb{P}[u \mid t, q] \propto |u_t| / \sum_{v \in \mathcal{X}} |v_t|$.

We then use the resulting distributions to sample data points as before, but every time a data point u is sampled, instead of incrementing its count in the histogram by one, we add $\textsc{Sign}(q_t u_t)$ to its entry. As the following lemma shows, in expectation, the final count is proportional to $\langle q, u \rangle$.

Lemma 8.1 *Define the random variable Z as 0 if data point $u \in \mathcal{X}$ is not sampled and $\textsc{Sign}(q_t u_t)$ if it is for a query $q \in \mathbb{R}^d$ and a sampled dimension t. Then $\mathbb{E}[Z] = \langle q, u \rangle / \sum_{t=1}^{d} \sum_{v \in \mathcal{X}} |q_t v_t|$.*

Proof.

$$
\begin{aligned}
\mathbb{E}[Z \mid t] &= \textsc{Sign}(q_t u_t)\, \mathbb{P}[u \mid t] = \textsc{Sign}(q_t u_t) \frac{|u_t|}{\sum_{v \in \mathcal{X}} |v_t|} \\
&= \textsc{Sign}(q_t u_t) \frac{|q_t u_t|}{\sum_{v \in \mathcal{X}} |q_t v_t|} = \frac{q_t u_t}{\sum_{v \in \mathcal{X}} |q_t v_t|}.
\end{aligned}
$$

Taking expectation over the dimension t yields:

$$
\begin{aligned}
\mathbb{E}[Z] = \mathbb{E}\Big[\mathbb{E}[Z \mid t] \Big] &= \sum_{t=1}^{d} \frac{q_t u_t}{\sum_{v \in \mathcal{X}} |q_t v_t|} \mathbb{P}[t \mid q] \\
&= \sum_{t=1}^{d} \frac{q_t u_t}{\sum_{v \in \mathcal{X}} |q_t v_t|} \frac{\sum_{v \in \mathcal{X}} |q_t v_t|}{\sum_{l=1}^{d} \sum_{v \in \mathcal{X}} |q_l v_l|} \\
&= \frac{\langle q, u \rangle}{\sum_{t=1}^{d} \sum_{v \in \mathcal{X}} |q_t v_t|}.
\end{aligned}
$$

□

8.2.3 Sample Complexity

We have formalized an efficient way to sample data points according to the distribution of inner products, and subsequently collect the most frequently-sampled points. But how many samples must we draw in order to accurately identify the top-k solution set? Ding et al. [2019] give an answer in the form of the following theorem for top-1 MIPS.

Before stating the result, it would be helpful to introduce a few shorthands. Let $N = \sum_{t=1}^{d} \sum_{v \in \mathcal{X}} |q_t v_t|$ be a normalizing factor. For a vector $u \in \mathcal{X}$, denote by Δ_u the scaled gap between the maximum inner product and the inner product of u and q: $\Delta_u = \langle q, u^* - u \rangle / N$.

If S is the number of samples to be drawn, for a vector u, denote by $Z_{u,i}$ a random variable that is 0 if u was not sampled in round i, and otherwise $\textsc{Sign}(q_t u_t)$ if t is the sampled dimension. Once the sampling has concluded, the final value for point u is simply $Z_u = \sum_i Z_{u,i}$. Note that, from Lemma 8.1, we have that $\mathbb{E}[Z_{u,i}] = \langle q, u \rangle / N$.

Given the notation above, let us also introduce the following helpful lemma.

Lemma 8.2 *Let $C_u = \sum_{t=1}^{d} |q_t u_t|$ for a data point u. Then for a pair of distinct vectors $u, v \in \mathcal{X}$:*

$$\mathbb{E}\Big[\big(Z_{u,i} - Z_{v,i}\big)^2\Big] = \frac{C_u + C_v}{N},$$

and,

$$\mathrm{Var}\Big[Z_u - Z_v\Big] = S\Big[\frac{C_u + C_v}{N} - \frac{\langle q, u - v \rangle^2}{N^2}\Big].$$

Proof. The proof is similar to the proof of Lemma 8.1. □

Theorem 8.1 *Suppose u^* is the exact solution to MIPS over m points in $\mathcal{X}$ for query q. Define $\sigma_u^2 = \mathrm{Var}\Big[Z_{u^*} - Z_u\Big]$ and let $\Delta = \min_{u \in \mathcal{X}} \Delta_u$. For $\delta \in (0, 1)$, if we drew S samples such that:*

$$S \geq \max_{u \neq u^*} \frac{(1 + \Delta_u)^2}{\sigma_u^2 h(\frac{\Delta_u (1 + \Delta_u)}{\sigma_u^2})} \log \frac{m}{\delta},$$

where $h(x) = (1 + x)\log(1 + x) - x$, then

$$\mathbb{P}[Z_{u^*} > Z_u \quad \forall\, u \neq u^*] \geq 1 - \delta.$$

Before proving the theorem above, let us make a quick observation. Clearly $\sigma_u^2 \leq \mathcal{O}(d\Delta_u)$ and $(1 + \Delta_u) \approx 1$. Because $h(\cdot)$ is monotone increasing in its argument ($\frac{\partial h}{\partial x} > 0$), we can write:

$$\sigma_u^2 h(\frac{\Delta_u(1+\Delta_u)}{\sigma_u^2}) = (\sigma_u^2 + \Delta_u(1+\Delta_u)) \log(1 + \frac{\Delta_u(1+\Delta_u)}{\sigma_u^2}) - \Delta_u(1+\Delta_u)$$
$$\approx \frac{\Delta_u^2(1+\Delta_u)^2}{\sigma_u^2} \geq \frac{\Delta}{d}(1+\Delta)^2 = \mathcal{O}(\frac{\Delta}{d}).$$

Plugging this into Theorem 8.1 gives us $S \leq \mathcal{O}(\frac{d}{\Delta} \log \frac{m}{\delta})$.

Theorem 8.1 tells us that, if we draw $\mathcal{O}(\frac{d}{\Delta} \log \frac{m}{\delta})$ samples, we can identify the top-1 solution to MIPS with high probability. Observe that, Δ is a measure of the difficulty of the query: When inner products are close to each other, Δ becomes smaller, implying that a larger number of samples would be needed to correctly identify the exact solution.

Proof of Theorem 8.1. Consider the probability that the registered value of a data point u is greater than or equal to the registered value of the solution u^* once sampling has concluded. That is, $\mathbb{P}[Z_u \geq Z_{u^*}]$. Let us rewrite that quantity as follows:

$$\mathbb{P}\Big[Z_u \geq Z_{u^*}\Big] = \mathbb{P}\Big[\sum_i Z_{u,i} - Z_{u^*,i} \geq 0\Big]$$
$$= \mathbb{P}\Big[\sum_{i=1}^{S} \underbrace{Z_{u,i} - Z_{u^*,i} + \Delta_u}_{Y_{u,i}} \geq \underbrace{S\Delta_u}_{y_u}\Big].$$

Notice that $\mathbb{E}[Y_{u,i}] = 0$ and that $Y_{u,i}$'s are independent. Furthermore, $Y_{u,i} \leq 1 + \Delta_u$. Letting $Y_u = \sum_i Y_{u,i}$, we can apply Bennett's inequality to bound the probability above:

$$\mathbb{P}[Y_u \geq y_u] \leq \exp\Bigg(-\frac{S\sigma_u^2}{(1+\Delta_u)^2} h\Big(\frac{(1+\Delta_u)(S\Delta_u)}{S\sigma_u^2}\Big)\Bigg).$$

Setting the right-hand-side to $\frac{\delta}{m}$, we arrive at:

$$\exp\Bigg(-\frac{S\sigma_u^2}{(1+\Delta_u)^2} h\Big(\frac{(1+\Delta_u)\Delta_u}{\sigma_u^2}\Big)\Bigg) \leq \frac{\delta}{m}$$
$$\implies S(1+\Delta_u)^{-2}\sigma_u^2 h\Big(\frac{\Delta_u(1+\Delta_u)}{\sigma_u^2}\Big) \geq \log\frac{m}{\delta}.$$

It is easy to see that for $x > 0$, $h(x) > 0$. Observing that $\Delta_u(1+\Delta_u)/\sigma_u^2$ is positive, that implies that $h(\Delta_u(1+\Delta_u)/\sigma_u^2) > 0$, and therefore we can re-arrange the expression above as follows:

$$S \geq \frac{(1+\Delta_u)^2}{\sigma_u^2 h\Big(\frac{\Delta_u(1+\Delta_u)}{\sigma_u^2}\Big)} \log \frac{m}{\delta}. \tag{8.4}$$

We have thus far shown that when S satisfies the inequality in (8.4), then $\mathbb{P}[Y_u \geq y_u] \leq \frac{\delta}{m}$. Going back to the claim, we derive the following bound using the result above:

$$\begin{aligned} &\mathbb{P}[Z_{u^*} > Z_u \quad \forall u \in \mathcal{X}] \\ &\quad = 1 - \mathbb{P}[\exists\, u \in \mathcal{X} \; s.t. \quad Z_{u^*} \leq Z_u] \\ &\quad \geq 1 - m\frac{\delta}{m} = 1 - \delta, \end{aligned}$$

where we have used the union bound to obtain the inequality. □

8.3 Approximating the Scores

The method we have just presented avoids the computation of inner products altogether but estimates the *rank* of each data point with respect to a query using a sampling procedure. In this section, we introduce another sampling method that approximates the *inner product* of every data point instead.

Let us motivate our next algorithm with a rather contrived example. Suppose that our data points and queries are in $\mathbb{R}^2$, with the first coordinate of vectors drawing values from $\mathcal{N}(0, \sigma_1^2)$ and the second coordinate from $\mathcal{N}(0, \sigma_2^2)$. If we were to compute the inner product of q with every vector $u \in \mathcal{X}$, we would need to perform two multiplications and a sum: $u_1 q_1 + u_2 q_2$. That gives us the exact "score" of every point with respect to q. But if $\sigma_1^2 \gg \sigma_2^2$, then by computing $q_1 u_1$ for all $u \in \mathcal{X}$, it is very likely that we have a good approximation to the final inner product. So we may use the partial inner product as a high-confidence estimate of the full inner product.

That is the core idea in this section. For each data point, we sample a few dimensions without replacement, and compute its partial inner product with the query along the chosen dimensions. Based on the scores so far, we can eliminate data points whose full inner product is projected, with high confidence, to be too small to make it to the top-k set. We then repeat the procedure by sampling more dimensions for the remaining data points, until we reach a stopping criterion.

The process above saves us time by shrinking the set of data points and computing only partial inner products in each round. But we must decide how we should sample dimensions and how we should determine which data points to discard. The objective is to minimize the number of samples needed to identify the solution set. These are the questions that Liu et al. [2019] answered in their work, which we will review next. We note that, even though Liu et al. [2019] use the Bandit language [Lattimore and Szepesvári, 2020] to describe

Algorithm 4: The BoundedME algorithm for MIPS.

Input: Query point $q \in \mathbb{R}^d$; $k \geq 1$ for top-k retrieval; confidence parameters $\epsilon, \delta \in (0, 1)$; and data points $\mathcal{X} \subset \mathbb{R}^d$

Result: $(1 - \delta)$-confident ϵ-approximate top-k set to MIPS with respect to q.

1: $i \leftarrow 1$

2: $\mathcal{X}_i \leftarrow \mathcal{X}$; ▷ `Initialize the solution set to` $\mathcal{X}$.

3: $\epsilon_i \leftarrow \frac{\epsilon}{4}$ and $\delta_i \leftarrow \frac{\delta}{2}$

4: $A_u \leftarrow 0 \quad \forall\, u \in \mathcal{X}_i$; ▷ A `is a score accumulator.`

5: $t_0 \leftarrow 0$

6: **while** $|\mathcal{X}_i| > k$ **do**

7: $\quad t_i \leftarrow h\left(\frac{2}{\epsilon_i^2} \log\left(\frac{2(|\mathcal{X}_i| - k)}{\delta_i(\lfloor \frac{|\mathcal{X}_i| - k}{2} \rfloor + 1)}\right)\right)$

8: $\quad$ **for** $u \in \mathcal{X}_i$ **do**

9: $\quad\quad$ Let $\mathcal{J}$ be $(t_i - t_{i-1})$ dimensions sampled without replacement

10: $\quad\quad A_u \leftarrow A_u + \sum_{j \in \mathcal{J}} u_j q_j$; ▷ `Compute partial inner product.`

11: $\quad$ **end for**

12: $\quad$ Let α be the $\lceil \frac{|\mathcal{X}_i| - k}{2} \rceil$-th score in A

13: $\quad \mathcal{X}_{i+1} \leftarrow \{u \in \mathcal{X}_i \; s.t. \; A_u > \alpha\}$

14: $\quad \epsilon_{i+1} \leftarrow \frac{3}{4}\epsilon_i$, $\delta_{i+1} \leftarrow \frac{\delta_i}{2}$, and $i \leftarrow i + 1$

15: **end while**

16: **return** X_i

their algorithm, we find it makes for a clearer presentation if we avoided the Bandit terminology.

8.3.1 The BoundedME Algorithm

The top-k retrieval algorithm developed by Liu et al. [2019] is presented in Algorithm 4. It is important to note that, for the algorithm to be correct—as we will explain later—each partial inner product must be bounded. In other words, for query q, any data point $u \in \mathcal{X}$, and any dimension t, we must have that $q_t u_t \in [a, b]$ for some fixed interval. This is not a restrictive assumption, however: q can always be normalized without affecting the solution to MIPS, and data points u can be scaled into the hypercube. In their work, Liu et al. [2019] assume that partial inner products are in the unit interval.

This iterative algorithm begins with the full collection of data points and removes almost half of the data points in each iteration. It terminates as soon as the total number of data points left is at most k.

In each iteration of the algorithm, it accumulates partial inner products for all remaining data point along a set of sampled dimensions. Once a dimension has been sampled, it is removed from consideration in all future iterations—hence, sampling *without* replacement.

The number of dimensions to sample is adaptive and changes from iteration to iteration. It is determined using the quantity on Line 7 of the algorithm, where the function $h(\cdot)$ is defined as follows:

$$h(x) = \min\Big\{\frac{1+x}{1+x/d}, \frac{x+x/d}{1+x/d}\Big\}. \tag{8.5}$$

At the end of iteration i with the remaining data points in $\mathcal{X}_i$, the algorithm finds the $\lceil \frac{|\mathcal{X}_i|-k}{2} \rceil$-th (i.e., close to the median) partial inner product accumulated so far, and discards data points whose score is less than that threshold. It then updates the confidence parameters ϵ and δ, and proceeds to the next iteration.

It is rather obvious that, the total number of dimensions along which the algorithm computes partial inner products for any given data point can never exceed d. That is simply because once Line 10 is executed, the dimensions in the set $\mathcal{J}$ defined on Line 9 are never considered for sampling in future iterations. As a result, in the worst case, the algorithm computes full inner products in $\mathcal{O}(md)$ operations.

As for the time complexity of Algorithm 4, it can be shown that it requires $\mathcal{O}(\frac{m\sqrt{d}}{\epsilon}\sqrt{\log(1/\delta)})$ operations. That is simply due to the fact that in each iteration, the number of data points is cut in half, combined with the inequality $h(x) \leq \mathcal{O}(\sqrt{dx})$ for $x > 0$.

Theorem 8.2 *The time complexity of Algorithm 4 is* $\mathcal{O}(\frac{m\sqrt{d}}{\epsilon}\sqrt{\log(1/\delta)})$.

Theorem 8.2 says that the time complexity of Algorithm 4 is linear in the number of data points m, but sub-linear in the number of dimensions d. That is a fundamentally different behavior than all the other algorithms we have presented thus far throughout the preceding chapters.

Proof of Theorem 8.2. Let us first show the following claim: $h(x) \leq \mathcal{O}(\sqrt{dx})$ for $x > 0$. To prove that, observe that $h(x)$ is the minimum of two positive values a and b. As such, $h(x) \leq \sqrt{ab}$. Substituting a and b with the right expressions from Equation (8.5):

$$\begin{aligned}
h(x) &\leq \sqrt{\frac{1+x}{1+x/d}\frac{x+x/d}{1+x/d}} = \frac{1}{1+x/d}\sqrt{x(1+x)(1+1/d)} \\
&= \frac{\mathcal{O}(x)}{1+x/d} = \mathcal{O}(\frac{dx}{d+x}) \leq \mathcal{O}(\sqrt{dx}).
\end{aligned}$$

Note that, in the i-th iteration there are at most $m/2^i$ data points to examine. Moreover, for each data point that is eliminated in round i, we will have computed at most t_i partial inner products (see Line 7 of Algorithm 4). Using these facts, we can calculate the time complexity as follows:

$$\sum_{i=1}^{\log m} \frac{m}{2^i} h(t_i) \leq \sum_{i=1}^{\log m} \frac{m}{2^i} \sqrt{dt_i} \leq \mathcal{O}\Big(\frac{m\sqrt{d}}{\epsilon}\sqrt{\log \frac{1}{\delta}}\Big).$$

□

8.3.2 Proof of Correctness

Our goal in this section is to prove that Algorithm 4 is correct, in the sense that it returns the ϵ-approximate solution to k-MIPS with probability at least $1-\delta$:

Theorem 8.3 *Algorithm 4 is guaranteed to return the ϵ-approximate solution to k-MIPS with probability at least $1-\delta$.*

The proof of Theorem 8.3 requires the concentration inequality due to Bardenet and Maillard [2015], repeated below for completeness.

Lemma 8.3 *Let $\mathcal{J} \subset [0,1]$ be a finite set of size d with mean μ. Let $\{J_1, J_2, \ldots, J_n\}$ be $n < d$ samples from $\mathcal{J}$ without replacement. Then for any $n \leq d$ and any $\delta \in [0,1]$ it holds:*

$$\mathbb{P}\Big[\frac{1}{n}\sum_{t=1}^{n} J_t - \mu \leq \sqrt{\frac{\rho_n}{2n}\log\frac{1}{\delta}}\Big] \geq 1-\delta,$$

where ρ_n is defined as follows:

$$\rho_n = \min\Big\{1 - \frac{n-1}{d}, (1-\frac{n}{d})(1+\frac{1}{n})\Big\}.$$

The lemma above guarantees that, with probability at least $1-\delta$, the empirical mean of the samples does not exceed the mean of the universe by a specific amount that depends on δ. We now wish to adapt that result to derive a similar guarantee where the difference between means is bounded by an arbitrary parameter ϵ. That is stated in the following lemma.

Lemma 8.4 *Let $\mathcal{J} \subset [0,1]$ be a finite set of size d with mean μ. Let $\{J_1, J_2, \ldots, J_n\}$ be $n < d$ samples from $\mathcal{J}$ without replacement. Then for any $\epsilon, \delta \in (0,1)$, if we have that:*

$$n \geq \min\Big\{\frac{1+x}{1+x/d}, \frac{x+x/d}{1+x/d}\Big\},$$

where $x = \log(1/\delta)/2\epsilon^2$, then the following holds:

$$\mathbb{P}\Big[\frac{1}{n}\sum_{t=1}^{n} J_t - \mu \leq \epsilon\Big] \geq 1-\delta.$$

Proof. By Lemma 8.3 we can see that:

$$\mathbb{P}\left[\frac{1}{n}\sum_{t=1}^{n} J_t - \mu \leq \epsilon\right] \geq 1 - \delta,$$

so long as:

$$\sqrt{\frac{\rho_n}{2n}\log\frac{1}{\delta}} \leq \epsilon \implies \frac{n}{\rho_n} \geq \frac{1}{2\epsilon^2}\log\frac{1}{\delta}.$$

There are two cases to consider. First, if $\rho_n = 1 - (n-1)/d$, then:

$$\frac{n}{\rho_n} \geq \underbrace{\frac{1}{2\epsilon^2}\log\frac{1}{\delta}}_{x} \implies \frac{n}{1 - \frac{n-1}{d}} \geq x$$
$$\implies n \geq \frac{x + x/d}{1 + x/d}.$$

In the second case, $\rho_n = (1 - n/d)(1 + 1/n)$, which gives:

$$\frac{n}{\rho_n} \geq \underbrace{\frac{1}{2\epsilon^2}\log\frac{1}{\delta}}_{x} \implies \frac{n}{(1 - \frac{n}{d})(1 + \frac{1}{n})} \geq x$$
$$\implies n \geq \left[1 + \frac{1}{n} - \frac{n+1}{d}\right]x$$
$$\implies n^2 \geq nx + x - \frac{n^2}{d}x - \frac{n}{d}x$$
$$\implies (1 + \frac{x}{d})n^2 - (x - \frac{x}{d})n - x \geq 0.$$

To make the closed-form solution more manageable, Liu et al. [2019] relax the problem above and solve n in the following problem instead. Note that, any solution to the problem below is a valid solution to the problem above.

$$(1 + \frac{x}{d})n^2 - (x - \frac{x}{d})n - x - 1 \geq 0 \implies \left[(1 + \frac{x}{d})n - x - 1\right][n + 1] \geq 0$$
$$\implies n \geq \frac{1 + x}{1 + x/d}.$$

By combining the two cases, we obtain:

$$n \geq \min\{\frac{1 + x}{1 + x/d}, \frac{x + x/d}{1 + x/d}\}.$$

□

Lemma 8.4 gives us the minimum number of dimensions we must sample so that the partial inner product of a vector with a query is at most ϵ away from the full inner product, with probability at least $1 - \delta$.

Armed with this result, we can now proceed to proving the main theorem.

Proof of Theorem 8.3. Denote by ζ_i the k-th largest *full* inner product among the set of data points $\mathcal{X}_i$ in iteration i. If we showed that, for two consecutive iterations, the difference between ζ_i and ζ_{i+1} does not exceed ϵ_i with probability at least $1 - \delta_i$, that is:

$$\mathbb{P}\Big[\zeta_i - \zeta_{i+1} \leq \epsilon_i\Big] \geq 1 - \delta_i, \tag{8.6}$$

then the theorem immediately follows:

$$\mathbb{P}\Big[\zeta_1 - \zeta_{\log m} \leq \epsilon\Big] \geq 1 - \delta,$$

because:

$$\sum_{i=1}^{\log m} \delta_i = \sum_{i=1}^{\log m} \frac{\delta}{2^i} \leq \sum_{i=1}^{\infty} \frac{\delta}{2^i} = \delta,$$

and,

$$\sum_{i=1}^{\log m} \epsilon_i = \sum_{i=1}^{\log m} \frac{\epsilon}{4}\Big(\frac{3}{4}\Big)^{i-1} \leq \sum_{i=1}^{\infty} \frac{\epsilon}{4}\Big(\frac{3}{4}\Big)^{i-1} = \epsilon.$$

So we focus on proving Equation (8.6).

Suppose we are in the i-th iteration. Collect in $\mathcal{Z}_{\epsilon_i}$ every data point in $u \in \mathcal{X}_i$ such that $\zeta_i - \langle q, u \rangle \leq \epsilon_i$. That is: $\mathcal{Z}_{\epsilon_i} = \{u \in \mathcal{X}_i \mid \zeta_i - \langle q, u \rangle \leq \epsilon_i\}$. If at least k elements of $\mathcal{Z}_{\epsilon_i}$ end up in $\mathcal{X}_{i+1}$, the event $\zeta_i - \zeta_{i+1} \leq \epsilon_i$ succeeds. So, that event fails if there are more than $\lfloor \frac{|\mathcal{X}_i| - k}{2} \rfloor$ data points in $\mathcal{X}_i \setminus \mathcal{Z}_{\epsilon_i}$ with partial inner products that are greater than partial inner products of the data points in $\mathcal{Z}_{\epsilon_i}$. Denote the number of such data points by β.

What is the probability that a data point u in $\mathcal{X}_i \setminus \mathcal{Z}_{\epsilon_i}$ has a higher partial inner product than any data point in $\mathcal{Z}_{\epsilon_i}$? Assuming that u^* is the data point that achieves ζ_i, we can write:

$$\begin{aligned}
\mathbb{P}\big[A_u \geq A_v \quad \forall\, v \in \mathcal{Z}_{\epsilon_i}\big] &\leq \mathbb{P}\big[A_u \geq A_{u^*}\big] \\
&\leq \mathbb{P}\big[A_u \geq \langle q, u \rangle + \frac{\epsilon_i}{2} \;\vee\; A_{u^*} \leq \zeta_i - \frac{\epsilon_i}{2}\big] \\
&\leq \mathbb{P}\big[A_u \geq \langle q, u \rangle + \frac{\epsilon_i}{2}\big] + \mathbb{P}\big[A_{u^*} \leq \zeta_i - \frac{\epsilon_i}{2}\big].
\end{aligned}$$

We can apply Lemma 8.4 to obtain that, if the number of sampled dimensions is equal to the quantity on Line 7 of Algorithm 4, then the probability above would be bounded by:

$$\frac{\lfloor \frac{|\mathcal{X}_i|-k}{2} \rfloor + 1}{|\mathcal{X}_i| - k} \delta_i.$$

Using this result along with Markov's inequality, we can bound the probability that β is strictly greater than $\lfloor \frac{|\mathcal{X}_i|-k}{2} \rfloor$ as follows:

$$\begin{aligned}
\mathbb{P}\left[\beta \geq \frac{|\mathcal{X}_i| - k}{2} + 1\right] &\leq \frac{\mathbb{E}[\beta]}{\frac{|\mathcal{X}_i|-k}{2} + 1} \\
&\leq \frac{(|\mathcal{X}_i| - k)\frac{\lfloor \frac{|\mathcal{X}_i|-k}{2} \rfloor + 1}{|\mathcal{X}_i|-k}\delta_i}{\frac{|\mathcal{X}_i|-k}{2} + 1} \\
&= \delta_i.
\end{aligned}$$

That completes the proof of Equation (8.6) and, therefore, the theorem. □

8.4 Closing Remarks

The algorithms in this chapter were unique in two ways. First, they directly took on the challenging problem of MIPS. This is in contrast to earlier chapters where MIPS was only an afterthought. Second, there is little to no preprocessing involved in the preparation of the index, which itself is small in size. That is unlike trees, hash buckets, graphs, and clustering that require a generally heavy index that itself is computationally-intensive to build.

The approach itself is rather unique as well. It is particularly interesting because the trade-off between efficiency and accuracy can be adjusted during retrieval. That is not the case with trees, LSH, or graphs, where the construction of the index itself heavily influences that balance. With sampling methods, it is at least theoretically possible to adapt the retrieval strategy to the hardness of the query distribution. That question remains unexplored.

Another area that would benefit from further research is the sampling strategy itself. In particular, in the BoundedME algorithm, the dimensions that are sampled next are drawn randomly. While that simplifies analysis—which follows the analysis of popular Bandit algorithms—it is not hard to argue that the strategy is sub-optimal. After all, unlike the Bandit setup, where reward distributions are unknown and samples from the reward distributions are revealed only gradually, here we have direct access to all data points *a priori*. Whether and how adapting the sampling strategy to the underlying data or query distribution may improve the error bounds or the accuracy or efficiency of the algorithm in practice remains to be studied.

References

G. Ballard, T. G. Kolda, A. Pinar, and C. Seshadhri. Diamond sampling for approximate maximum all-pairs dot-product (mad) search. In *2015 IEEE International Conference on Data Mining*, pages 11–20, 2015.

R. Bardenet and O.-A. Maillard. Concentration inequalities for sampling without replacement. *Bernoulli*, 21(3):1361–1385, 2015.

E. Cohen and D. D. Lewis. Approximating matrix multiplication for pattern recognition tasks. In *Proceedings of the Eighth Annual ACM-SIAM Symposium on Discrete Algorithms*, pages 682–691, 1997.

Q. Ding, H.-F. Yu, and C.-J. Hsieh. A fast sampling algorithm for maximum inner product search. In K. Chaudhuri and M. Sugiyama, editors, *Proceedings of the 22nd International Conference on Artificial Intelligence and Statistics*, volume 89 of *Proceedings of Machine Learning Research*, pages 3004–3012, 4 2019.

T. Lattimore and C. Szepesvári. *Bandit Algorithms.* Cambridge University Press, 2020.

R. Liu, T. Wu, and B. Mozafari. A bandit approach to maximum inner product search. In *Proceedings of the 33rd AAAI Conference on Artificial Intelligence*, 2019.

S. S. Lorenzen and N. Pham. Revisiting wedge sampling for budgeted maximum inner product search. In *Proceedings of the 30th International Joint Conference on Artificial Intelligence*, pages 4789–4793, 8 2021.

A. J. Walker. An efficient method for generating discrete random variables with general distributions. *ACM Transactions on Mathematical Software*, 3(3):253–256, 9 1977.

Part III
Compression

Chapter 9
Quantization

Abstract In a vector retrieval system, it is usually not enough to process queries as fast as possible. It is equally as important to reduce the size of the index by compressing vectors. Compression, however, must be done in such a way that either decompressing the vectors during retrieval incurs a negligible cost, or distances can be computed (approximately) in the compressed domain, rendering it unnecessary to decompress compressed vectors during retrieval. This chapter introduces a class of vector compression algorithms, known as quantization, that is inspired by clustering.

9.1 Vector Quantization

Let us take a step back and present a different mental model of the clustering-based retrieval framework discussed in Chapter 7. At a high level, we band together points that are placed by $\zeta(\cdot)$ into cluster i and represent that group by μ_i, for $i \in [C]$. In the first stage of the search for query q, we take the following conceptual step: First, we compute $\delta(q, \mu_i)$ for every i and construct a "table" that maps i to $\delta(q, \mu_i)$. We next approximate $\delta(q, u)$ for every $u \in \mathcal{X}$ using the resulting table: If $u \in \zeta^{-1}(i)$, then we look up an estimate of its distance to q from the i-th row of the table. We then identify the ℓ closest distances, and perform a secondary search over the corresponding vectors.

This presentation of clustering for top-k retrieval highlights an important fact that does not come across as clearly in our original description of the algorithm: We have made an implicit assumption that $\delta(q, u) \approx \delta(q, \mu_i)$ for all $u \in \zeta^{-1}(i)$. That is why we presume that if a cluster minimizes $\delta(q, \cdot)$, then the points within it are also likely to minimize $\delta(q, \cdot)$. That is, in turn, why we deem it sufficient to search over the points within the top-ℓ clusters.

Put differently, within the first stage of search, we appear to be approximating every point $u \in \zeta^{-1}(i)$ with $\tilde{u} = \mu_i$. Because there are C discrete choices to consider for every data point, we can say that we *quantize* the

S. Bruch, *Foundations of Vector Retrieval*, https://doi.org/10.1007/978-3-031-55182-6_9

vectors into $[C]$. Consequently, we can encode each vector using only $\log_2 C$ bits, and an entire collection of vectors using $m \log_2 C$ bits! All together, we can represent a collection $\mathcal{X}$ using $\mathcal{O}(Cd + m \log_2 C)$ space, and compute distances to a query by performing m look-ups into a table that itself takes $\mathcal{O}(Cd)$ time to construct. That quantity can be far smaller than $\mathcal{O}(md)$ given by the naïve distance computation algorithm.

Clearly, the approximation error, $\|u - \tilde{u}\|$, is a function of C. As we increase C, this approximation improves, so that $\|u - \tilde{u}\| \rightarrow 0$ and $|\delta(q,u) - \delta(q,\tilde{u})| \rightarrow 0$. Indeed, $C = m$ implies that $\tilde{u} = u$ for every u. But increasing C results in an increased space complexity and a less efficient distance computation. At $C = m$, for example, our table-building exercise does not help speed up distance computation for individual data points—because we must construct the table in $\mathcal{O}(md)$ time anyway. Finding the right C is therefore critical to space- and time-complexity, as well as the approximation or quantization error.

9.1.1 Codebooks and Codewords

What we described above is known as vector quantization [Gray and Neuhoff, 1998] for vectors in the L_2 space. We will therefore assume that $\delta(u,v) = \|u - v\|_2^2$ in the remainder of this section. The function $\zeta : \mathbb{R}^d \rightarrow [C]$ is called a *quantizer*, the individual centroids are referred to as *codewords*, and the set of C codewords make up a *codebook*. It is easy to see that the set $\zeta^{-1}(i)$ is the intersection of $\mathcal{X}$ with the Voronoi region associated with codeword μ_i.

The approximation quality of a given codebook is measured by the familiar mean squared error: $\mathbb{E}[\|\mu_{\zeta(U)} - U\|_2^2)]$, with U denoting a random vector. Interestingly, that is exactly the objective that is minimized by Lloyd's algorithm for KMeans clustering. As such, an optimal codebook is one that satisfies Lloyd's optimality conditions: each data point must be quantized to its nearest codeword, and each Voronoi region must be represented by its mean. That is why KMeans is our default choice for ζ.

9.2 Product Quantization

As we noted earlier, the quantization error is a function of the number of clusters, C: A larger value of C drives down the approximation error, making the quantization and the subsequent top-k retrieval solution more accurate and effective. However, realistically, C cannot become too large, because then the framework would collapse to exhaustive search, degrading its efficiency. How may we reconcile the two seemingly opposing forces?

Jégou et al. [2011] gave an answer to that question in the form of Product Quantization (PQ). The idea is easy to describe at a high level: Whereas in vector quantization we quantize the entire vector into one of C clusters, in PQ we break up a vector into orthogonal subspaces and perform vector quantization on individual chunks separately. The quantized vector is then a concatenation of the quantized subspaces.

Formally, suppose that the number of dimensions d is divisible by $d_\circ$, and let $L = d/d_\circ$. Define a selector matrix $S_i \in \{0,1\}^{d_\circ \times d}$, $1 \leq i \leq L$ as a matrix with L blocks in $\{0,1\}^{d_\circ \times d_\circ}$, where all blocks are 0 but the i-th block is the identity. The following is an example for $d = 6$, $d_\circ = 2$, and $i = 2$:

$$S_2 = \begin{bmatrix} 0\,0\,1\,0\,0\,0 \\ 0\,0\,0\,1\,0\,0 \end{bmatrix}$$

For a given vector $u \in \mathbb{R}^d$, $S_i u$ gives the i-th $d_\circ$-dimensional subspace, so that we can write: $u = \bigoplus_i S_i u$. Suppose further that we have n quantizers ζ_1 through ζ_L, where $\zeta_i : \mathbb{R}^{d_\circ} \rightarrow [C]$ maps the subspace selected by S_i to one of C clusters. Each ζ_i gives us C centroids $\mu_{i,j}$ for $j \in [C]$.

Using the notation above, we can express the PQ code for a vector u as L cluster identifiers, $\zeta_i(S_i u)$, for $i \in [L]$. We can therefore quantize a d-dimensional vector using $L \log_2 C$ bits. Observe that, when $L = 1$ (or equivalently, $d_\circ = d$), PQ reduces to vector quantization. When $L = d$, on the other hand, PQ performs scalar quantization per dimension.

Given this scheme, our approximation of u is $\tilde{u} = \bigoplus_i \mu_{i,\zeta_i(u)}$. It is easy to see that the quantization error $\mathbb{E}[\lVert U - \tilde{U} \rVert_2^2]$, with U denoting a random vector drawn from $\mathcal{X}$ and $\tilde{U}$ its reconstruction, is the sum of the quantization error of individual subspaces:

$$\begin{aligned} \mathbb{E}[\lVert U - \tilde{U} \rVert_2^2] &= \frac{1}{m} \sum_{u \in \mathcal{X}} \Big[\lVert u - \bigoplus_{i=1}^{L} \mu_{i,\zeta_i(u)} \rVert_2^2 \Big] \\ &= \frac{1}{m} \sum_{u \in \mathcal{X}} \Big[\sum_{i=1}^{L} \lVert S_i u - \mu_{i,\zeta_i(u)} \rVert_2^2 \Big]. \end{aligned}$$

As a result, learning the L codebooks can be formulated as L independent sub-problems. The i-th codebook can therefore be learnt by the application of KMeans on $S_i \mathcal{X} = \{ S_i u \mid u \in \mathcal{X} \}$.

9.2.1 Distance Computation with PQ

In vector quantization, computing the distance of a vector u to a query q was fairly trivial. All we had to do was to precompute a table that maps $i \in [C]$ to $\lVert q - \mu_i \rVert_2$, then look up the entry that corresponds to $\zeta(u)$. The fact that we were able to precompute C distances once per query, then simply look up the right entry from the table for a vector u helped us save a great deal of computation. Can we devise a similar algorithm given a PQ code?

The answer is yes. Indeed, that is why PQ has proven to be an efficient algorithm for distance computation. As in vector quantization, it first computes L distance tables, but the i-th table maps $j \in [C]$ to $\|S_i q - \mu_{i,j}\|_2^2$ (note the *squared* L_2 distance). Using these tables, we can estimate the distance between q and any vector u as follows:

$$\begin{aligned}
\|q - u\|_2^2 &\approx \|q - \tilde{u}\|_2^2 \\
&= \|q - \bigoplus_{i=1}^{L} \mu_{i,\zeta_i(u)}\|_2^2 \\
&= \|\bigoplus_{i=1}^{L} \Big(S_i q - \mu_{i,\zeta_i(u)}\Big)\|_2^2 \\
&= \sum_{i=1}^{L} \|S_i q - \mu_{i,\zeta_i(u)}\|_2^2.
\end{aligned}$$

Observe that, we have already computed the summands and recorded them in the distance tables. As a result, approximating the distance between u and q amounts to L table look-ups. The overall amount of computation needed to approximate distances between q and m vectors in $\mathcal{X}$ is then $\mathcal{O}(LCd_\circ + mL)$.

We must remark on the newly-introduced parameter $d_\circ$. Even though in the context of vector quantization, the impact of C on the quantization error is not theoretically known, there is nonetheless a clear interpretation: A larger C leads to better quantization. In PQ, the impact of $d_\circ$ or, equivalently, L on the quantization error is not as clear. As noted earlier, we can say something about $d_\circ$ at the extremes, but what we should expect from a value somewhere between 1 and d is largely an empirical question [Sun et al., 2023].

9.2.2 Optimized Product Quantization

In PQ, we allocate an equal number of bits ($\log_2 C$) to each of the n orthogonal subspaces. This makes sense if our vectors have similar energy in every subspace. But when the dimensions in one subspace are highly correlated, and in another uncorrolated, our equal-bits-per-subspace allocation policy proves wasteful in the former and perhaps inadequate in the latter. How can we ensure a more balanced energy across subspaces?

Jégou et al. [2011] argue that applying a random rotation $R \in \mathbb{R}^{d \times d}$ ($RR^T = I$) to the data points prior to quantization is one way to reduce the correlation between dimensions. The matrix R together with S_i's, as defined above, determines how we decompose the vector space into its sub-spaces. By applying a rotation first, we no longer chunk up an input vector into sub-vectors that comprise of consecutive dimensions.

Later, Ge et al. [2014] and Norouzi and Fleet [2013] extended this idea and suggested that the matrix R can be learnt jointly with the codebooks. This can be done through an iterative algorithm that switches between two steps in each iteration. In the first step, we freeze R and learn a PQ codebook as before. In the second step, we freeze the codebook and update the matrix R by solving the following optimization problem:

$$\begin{aligned} \min_R \sum_{u \in \mathcal{X}} \|Ru - \tilde{u}\|_2^2, \\ s.t. \quad RR^T = I, \end{aligned}$$

where $\tilde{u}$ is the approximation of u according to the frozen PQ codebook. Because u and $\tilde{u}$ are fixed in the above optimization problem, we can rewrite the objective as follows:

$$\begin{aligned} \min_R \|RU - \tilde{U}\|_F, \\ s.t. \quad RR^T = I, \end{aligned}$$

where U is a d-by-m matrix where each column is a vector in $\mathcal{X}$, $\tilde{U}$ is a matrix where each column is an approximation of the corresponding column in U, and $\|\cdot\|_F$ is the Frobenius norm. This problem has a closed-form solution as shown by Ge et al. [2014].

9.2.3 Extensions

Since the study by Jégou et al. [2011], many variations of the idea have emerged in the literature. In the original publication, for example, Jégou et al. [2011] used PQ codes in conjunction with the clustering-based retrieval framework presented earlier in this chapter. In other words, a collection $\mathcal{X}$ is first clustered into C clusters ("coarse-quantization"), and each cluster is subsequently represented using its own PQ codebook. In this way, when the routing function identifies a cluster to search, we can compute distances for data points within that cluster using their PQ codes. Later, Babenko and Lempitsky [2012] extended this two-level quantization further by introducing the "inverted multi-index" structure.

When combining PQ with clustering or coarse-quantization, instead of producing PQ codebooks for raw vectors within each cluster, one could learn codebooks for the *residual* vectors instead. That means, if the centroid of the i-th cluster is μ_i, then we may quantize $(u - \mu_i)$ for each vector $u \in \zeta^{-1}(i)$.

This was the idea first introduced by Jégou et al. [2011], then developed further in subsequent works [Kalantidis and Avrithis, 2014, Wu et al., 2017].

The PQ literature does not end there. In fact, so popular, effective, and efficient is PQ that it pops up in many different contexts and a variety of applications. Research into improving its accuracy and speed is still ongoing. For example, there have been many works that speed up the distance computation with PQ codebooks by leveraging hardware capabilities [Johnson et al., 2021, Andre et al., 2021, André et al., 2015]. Others that extend the algorithm to streaming (online) collections [Xu et al., 2018], and yet other studies that investigate other PQ codebook-learning protocols [Liu et al., 2020, Yu et al., 2018, Chen et al., 2020, Jang and Cho, 2021, Klein and Wolf, 2019, Lu et al., 2023]. This list is certainly not exhaustive and is still growing.

9.3 Additive Quantization

PQ remains the dominant quantization method for top-k retrieval due to its overall simplicity and the efficiency of its codebook learning protocol. There are, however, numerous generalizations of the framework [Babenko and Lempitsky, 2014, Chen et al., 2010, Niu et al., 2023, Liu et al., 2015, Ozan et al., 2016, Krishnan and Liberty, 2021]. Typically, these generalized forms improve the approximation error but require more involved codebook learning algorithms and vector encoding protocols. In this section, we review one key algorithm, known as Additive Quantization (AQ) [Babenko and Lempitsky, 2014], that is the backbone of all other methods.

Like PQ, AQ learns L codebooks where each codebook consists of C codewords. Unlike PQ, however, each codeword is a vector in $\mathbb{R}^d$—rather than $\mathbb{R}^{d_\circ}$. Furthermore, a vector u is approximated as the *sum*, instead of the concatenation, of L codewords, one from each codebook: $\tilde{u} = \sum_{i=1}^{L} \mu_{i,\zeta_i(u)}$, where $\zeta_i : \mathbb{R}^d \rightarrow [C]$ is the quantizer associated with the i-th codebook.

Let us compare AQ with PQ at a high level and understand how AQ is different. We can still encode a data point using $L \log_2 C$ bits, as in PQ. However, the codebooks for AQ are L-times larger than their PQ counterparts, simply because each codeword has d dimensions instead of $d_\circ$. On the other hand, AQ does not decompose the space into orthogonal subspaces and, as such, makes no assumptions about the independence between subspaces.

AQ is therefore a strictly more general quantization method than PQ. In fact, the class of additive quantizers contains the class of product quantizers: By restricting the i-th codebook in AQ to the set of codewords that are 0 everywhere outside of the i-th "chunk," we recover PQ. Empirical comparisons [Babenko and Lempitsky, 2014, Matsui et al., 2018] confirm that such a generalization is more effective in practice.

For this formulation to be complete, we have to specify how the codebooks are learnt, how we encode an arbitrary vector, and how we perform distance computation. We will cover these topics in reverse order in the following sections.

9.3.1 Distance Computation with AQ

Suppose for the moment that we have learnt AQ codebooks for a collection $\mathcal{X}$ and that we are able to encode an arbitrary vector into an AQ code (i.e., a vector of L codeword identifiers). In this section, we examine how we may compute the distance between a query point q and a data point u using its approximation $\tilde{u}$.

Observe the following fact:

$$\|q - u\|_2^2 = \|q\|_2^2 - 2\langle q, u\rangle + \|u\|_2^2.$$

The first term is a constant that can be computed once per query and, at any rate, is inconsequential to the top-k retrieval problem. The last term, $\|u\|_2^2$ can be stored for every vector and looked up during distance computation, as suggested by Babenko and Lempitsky [2014]. That means, the encoding of a vector $u \in \mathcal{X}$ comprises of two components: $\tilde{u}$ and its (possibly scalar-quantized) squared norm. This brings the total space required to encode m vectors to $\mathcal{O}(LCd + m(1 + L\log_2 C))$.

The middle term can be approximated by $\langle q, \tilde{u}\rangle$ and can be expressed as follows:

$$\langle q, u\rangle \approx \langle q, \tilde{u}\rangle = \sum \langle q, \mu_{i,\zeta_i(u)}\rangle.$$

As in PQ, the summands can be computed once for all codewords, and stored in a table. When approximating the inner product, we can do as before and look up the appropriate entries from these precomputed tables. The time complexity of this operation is therefore $\mathcal{O}(LCd + mL)$ for m data points, which is similar to PQ.

9.3.2 AQ Encoding and Codebook Learning

While distance computation with AQ codes is fairly similar to the process involving PQ codes, the encoding of a data point is substantially different and relatively complex in AQ. That is because we can no longer simply assign a vector to its nearest codeword. Instead, we must find an arrangement of L codewords that together minimize the approximation error $\|u - \tilde{u}\|_2$.

Let us expand the expression for the approximation error as follows:

$$\begin{aligned}
\|u - \tilde{u}\|_2^2 &= \|u - \sum_{i=1}^{L} \mu_{i,\zeta_i(u)}\|_2^2 \\
&= \|u\|_2^2 - 2\langle u, \sum_{i=1}^{L} \mu_{i,\zeta_i(u)}\rangle + \|\sum_{i=1}^{L} \mu_{i,\zeta_i(u)}\|_2^2 \\
&= \|u\|_2^2 + \Big(\sum_{i=1}^{L} -2\langle u, \mu_{i,\zeta_i(u)}\rangle + \|\mu_{i,\zeta_i(u)}\|_2^2\Big) + \\
&\sum_{1 \leq i < j \leq L} 2\langle \mu_{i,\zeta_i(u)}, \mu_{j,\zeta_j(u)}\rangle.
\end{aligned}$$

Notice that the first term is irrelevant to the objective function, so we may ignore it. We must therefore find ζ_i's that minimize the remaining terms.

Babenko and Lempitsky [2014] use a generalized Beam search to solve this optimization problem. The algorithm begins by selecting L closest codewords from $\bigcup_{i=1}^{L}\{\mu_{i,1} \ldots \mu_{i,C}\}$ to u. For a chosen codeword $\mu_{k,j}$, we compute the residual $u - \mu_{k,j}$ and find the L closest codewords to it from $\bigcup_{i \neq k}\{\mu_{i,1} \ldots \mu_{i,C}\}$. After performing this search for all chosen codewords from the first round, we end up with a maximum of L^2 unique pairs of codewords. Note that, each pair has codewords from two different codebooks.

Of the L^2 pairs, the algorithm picks the top L that minimize the approximation error. It then repeats this process for a total of L rounds, where in each round we compute the residuals given L tuples of codewords, and for each tuple, find L codewords from the remaining codebooks, and ultimately identify the top L tuples from the L^2 tuples. At the end of the L-th round, the tuple with the minimal approximation error is the encoding for u.

Now that we have addressed the vector encoding part, it remains to describe the codebook learning procedure. Unsurprisingly, learning a codebook is not so dissimilar to the PQ codebook learning algorithm. It is an iterative procedure alternating between two steps to optimize the following objective:

$$\min_{\mu_{i,j}} \sum_{u \in \mathcal{X}} \|u - \sum_{i=1}^{L} \mu_{i,\zeta_i(u)}\|_2^2.$$

One step of every iteration freezes the codewords and performs assignments ζ_i's, which is the encoding problem we have already discussed above. The second step freezes the assignments and updates the codewords, which itself is a least-squares problem that can be solved relatively efficiently, considering that it decomposes over each dimension.

9.4 Quantization for Inner Product

The vector quantization literature has largely been focused on the Euclidean distance and the approximate nearest neighbor search problem. Those ideas typically port over to the maximum cosine similarity search with little effort, but not to MIPS under general conditions. To understand why, suppose we wish to find a quantizer such that the inner product approximation error is minimized for a query distribution:

$$\begin{aligned}\mathop{\mathbb{E}}_q\Big[\sum_{u\in\mathcal{X}}\big(\langle q,u\rangle-\langle q,\tilde{u}\rangle\big)^2\Big] &= \sum_{u\in\mathcal{X}}\mathop{\mathbb{E}}_q\Big[\langle q,u-\tilde{u}\rangle^2\Big] \\ &= \sum_{u\in\mathcal{X}}\mathop{\mathbb{E}}_q\Big[(u-\tilde{u})^T qq^T(u-\tilde{u})\Big] \\ &= \sum_{u\in\mathcal{X}}(u-\tilde{u})^T\mathop{\mathbb{E}}_q\big[qq^T\big](u-\tilde{u}), \qquad (9.1)\end{aligned}$$

where $\tilde{u}$ is an approximation of u. If we assumed that q is isotropic, so that its covariance matrix is the identity matrix scaled by some constant, then the objective above reduces to the reconstruction error. In that particular case, it makes sense for the quantization objective to be based on the reconstruction error, making the quantization methods we have studied thus far appropriate for MIPS too. But in the more general case, where the distribution of q is anisotropic, there is a gap between the true objective and the reconstruction error.

Guo et al. [2016] showed that, if we are able to obtain a small sample of queries to estimate $\mathbb{E}[qq^T]$, then we can modify the assignment step in Lloyd's iterative algorithm for KMeans in order to minimize the objective in Equation (9.1). That is, instead of assigning points to clusters by their Euclidean distance to the (frozen) centroids, we must instead use Mahalanobis distance characterized by $\mathbb{E}[qq^T]$. The resulting quantizer is arguably more suitable for inner product than the plain reconstruction error.

9.4.1 Score-aware Quantization

Later, Guo et al. [2020] argued that the objective in Equation (9.1) does not adequately capture the nuances of MIPS. Their argument rests on an observation and an intuition. The observation is that, in Equation (9.1), every single data point contributes equally to the optimization objective. Intuitively, however, data points are not equally likely to be the solution to MIPS. The error from data points that are more likely to be the maximizers of inner product with queries should therefore be weighted more heavily than others.

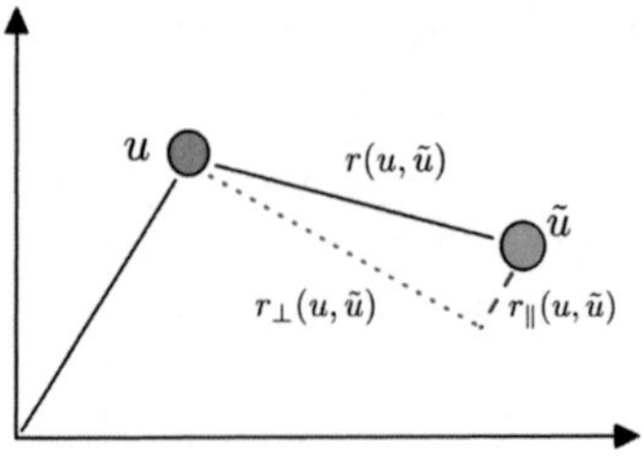

Fig. 9.1: Decomposition of the residual error $r(u, \tilde{u}) = u - \tilde{u}$ for $u \in \mathbb{R}^2$ to one component that is *parallel* to the data point, $r_\parallel(u, \tilde{u})$, and another that is *orthogonal* to it, $r_\perp(u, \tilde{u})$.

On the basis of that argument, Guo et al. [2020] introduce the following objective for inner product quantization:

$$\sum_{u \in \mathcal{X}} \underbrace{\mathbb{E}_q \left[\omega(\langle q, u \rangle) \, \langle q, u - \tilde{u} \rangle^2 \right]}_{\ell(u, \tilde{u}, \omega)}. \tag{9.2}$$

In the above, $\omega : \mathbb{R} \rightarrow \mathbb{R}^+$ is an arbitrary weight function that determines the importance of each data point to the optimization objective. Ideally, then, ω should be monotonically non-decreasing in its argument. One such weight function is $\omega(s) = \mathbb{1}_{s \geq \theta}$ for some threshold θ, implying that only data points whose expected inner product is at least θ contribute to the objective, while the rest are simply ignored. That is the weight function that Guo et al. [2020] choose in their work.

Something interesting emerges from Equation (9.2) with the choice of $\omega(s) = \mathbb{1}_{s \geq \theta}$: It is more important for $\tilde{u}$ to preserve the *norm* of u than it is to preserve its *angle*. We will show why that is shortly, but consider for the moment the reason this behavior is important for MIPS. Suppose there is a data point whose norm is much larger than the rest of the data points. Intuitively, such a data point has a good chance of maximizing inner product with a query even if its angle with the query is relatively large. In other words, being a candidate solution to MIPS is less sensitive to angles and more sensitive to norms. Of course, as norms become more and more concentrated, angles take on a bigger role in determining the solution to MIPS. So, intuitively, an objective that penalizes the distortion of norms more than angles is more suitable for MIPS.

9.4.1.1 Parallel and Orthogonal Residuals

Let us present this phenomenon more formally and show why the statement above is true. Define the residual error as $r(u, \tilde{u}) = u - \tilde{u}$. The residual error

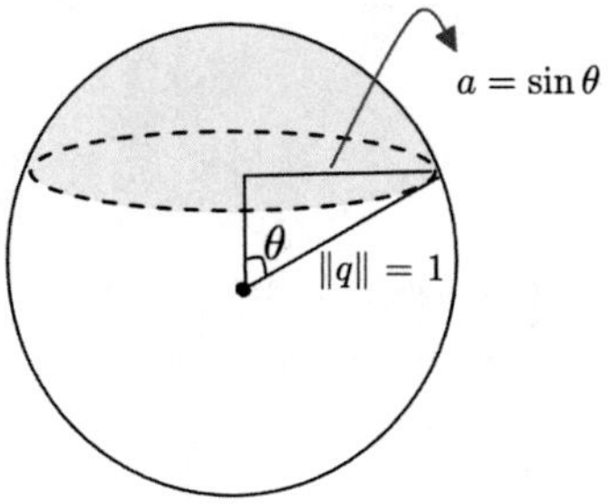

Fig. 9.2: The probability that the angle between a fixed data point u with a unit-normed query q that is drawn from a spherically-symmetric distribution is at most θ, is equal to the surface area of the spherical cap with base radius $a = \sin\theta$. This fact is used in the proof of Theorem 9.1.

can be decomposed into two components: one that is parallel to the data point, $r_{\|}(u, \tilde{u})$, and another that is orthogonal to it, $r_{\perp}(u, \tilde{u})$, as depicted in Figure 9.1. More concretely:

$$r_{\|}(u, \tilde{u}) = \frac{\langle u - \tilde{u}, u \rangle}{\|u\|^2} u,$$

and,

$$r_{\perp}(u, \tilde{u}) = r(u, \tilde{u}) - r_{\|}(u, \tilde{u}).$$

Guo et al. [2020] show first that, regardless of the choice of ω, the loss defined by $\ell(u, \tilde{u}, \omega)$ in Equation (9.2) can be decomposed as stated in the following theorem.

Theorem 9.1 *Given a data point u, its approximation $\tilde{u}$, and any weight function ω, the objective of Equation (9.2) can be decomposed as follows for a spherically-symmetric query distribution:*

$$\ell(u, \tilde{u}, \omega) \propto h_{\|}(\omega, \|u\|)\ \|r_{\|}(u, \tilde{u})\|^2 + h_{\perp}(\omega, \|u\|)\ \|r_{\perp}(u, \tilde{u})\|^2,$$

where,

$$h_{\|}(\omega, t) = \int_0^{\pi} \omega(t \cos\theta)\Big(\sin^{d-2}\theta - \sin^d\theta\Big)\ d\theta,$$

and,

$$h_{\perp}(\omega, t) = \frac{1}{d-1}\int_0^{\pi} \omega(t\cos\theta)\sin^d\theta\ d\theta.$$

Proof. Without loss of generality, we can assume that queries are unit vectors (i.e., $\|q\| = 1$). Let us write $\ell(u, \tilde{u}, \omega)$ as follows:

$$\ell(u, \tilde{u}, \omega) = \mathop{\mathbb{E}}_{q}\Big[\omega(\langle q, u\rangle)\ \langle q, u - \tilde{u}\rangle^2\Big]$$
$$= \int_0^\pi \omega(\lVert u\rVert \cos\theta) \mathop{\mathbb{E}}_{q}\Big[\langle q, u - \tilde{u}\rangle^2 \big\vert \langle q, u\rangle = \lVert u\rVert\cos\theta\Big]\ d\,\mathbb{P}\Big[\theta_{q,u} \leq \theta\Big],$$

where $\theta_{q,u}$ denotes the angle between q and u.

Observe that $\mathbb{P}\Big[\theta_{q,u} \leq \theta\Big]$ is the surface area of a spherical cap with base radius $a = \lVert q\rVert \sin\theta = \sin\theta$—see Figure 9.2. That quantity is equal to:

$$\lVert q\rVert^{d-1} \frac{\pi^{d/2}}{\Gamma(d/2)} I(a^2; \frac{d-1}{2}, \frac{1}{2}),$$

where Γ is the Gamma function and $I(z; \cdot, \cdot)$ is the incomplete Beta function. We may therefore write:

$$\frac{d\,\mathbb{P}\Big[\theta_{q,u} \leq \theta\Big]}{d\theta} \propto \Big[(1 - a^2)^{\frac{1}{2} - 1} (a^2)^{\frac{d-1}{2} - 1}\Big] \frac{da}{d\theta}$$
$$= \frac{\sin^{d-3}\theta}{\cos\theta}(2\sin\theta\cos\theta)$$
$$\propto \sin^{d-2}\theta,$$

where in the first step we used the fact that $dI(z; s, t) = (1 - z)^{t-1} z^{s-1} dz$.

Putting everything together, we can rewrite the loss as follows:

$$\ell(u, \tilde{u}, \omega) \propto \int_0^\pi \omega(\lVert u\rVert \cos\theta) \mathop{\mathbb{E}}_{q}\Big[\langle q, u - \tilde{u}\rangle^2 \big\vert \langle q, u\rangle = \lVert u\rVert\cos\theta\Big] \sin^{d-2}\theta\ d\theta.$$

We can complete the proof by applying the following lemma to the expectation over queries in the integral above.

Lemma 9.1

$$\mathop{\mathbb{E}}_{q}[\langle q, u - \tilde{u}\rangle^2 \vert \langle q, u\rangle = t] = \frac{t^2}{\lVert u\rVert^2} \lVert r_\parallel(u, \tilde{u})\rVert^2 + \frac{1 - t^2/\lVert u\rVert^2}{d - 1} \lVert r_\perp(u, \tilde{u})\rVert^2.$$

Proof. We use the shorthand $r_\parallel = r_\parallel(u, \tilde{u})$ and similarly $r_\perp = r_\perp(u, \tilde{u})$. Decompose $q = q_\parallel + q_\perp$ where $q_\parallel = \langle q, u\rangle \frac{u}{\lVert u\rVert^2}$ and $q_\perp = q - q_\parallel$. We can now write:

$$\mathop{\mathbb{E}}_{q}[\langle q, u - \tilde{u}\rangle^2 \vert \langle q, u\rangle = t] = \mathop{\mathbb{E}}_{q}[\langle q_\parallel, r_\parallel\rangle^2] \vert \langle q, u\rangle = t] + \mathop{\mathbb{E}}_{q}[\langle q_\perp, r_\perp\rangle^2] \vert \langle q, u\rangle = t].$$

All other terms are equal to 0 either due to orthogonality or components or because of spherical symmetry. The first term is simply equal to $\lVert r_\parallel\rVert^2 \frac{t^2}{\lVert u\rVert^2}$. By spherical symmetry, it is easy to show that the second term reduces to $\frac{1 - t^2/\lVert u\rVert^2}{d-1} \lVert r_\perp\rVert^2$. That completes the proof. □

Applying the lemma above to the integral, we obtain:

$$\ell(u, \tilde{u}, \omega) \propto \int_0^{\pi} \omega(\|u\| \cos\theta) \left(\cos^2\theta \|r_{\|}(u, \tilde{u})\|^2 + \frac{\sin^2\theta}{d-1} \|r_{\perp}(u, \tilde{u})\|^2 \right) \sin^{d-2}\theta \; d\theta,$$

as desired. □

When $\omega(s) = \mathbb{1}_{s \geq \theta}$ for some θ, Guo et al. [2020] show that $h_{\|}$ outweighs $h_{\perp}$, as the following theorem states. This implies that such an ω puts more emphasis on preserving the parallel residual error as discussed earlier.

Theorem 9.2 *For $\omega(s) = \mathbb{1}_{s \geq \theta}$ with $\theta \geq 0$, $h_{\|}(\omega, t) \geq h_{\perp}(\omega, t)$, with equality if and only if ω is constant over the interval $[-t, t]$.*

Proof. We can safely assume that $h_{\|}$ and $h_{\perp}$ are positive; they are 0 if and only if $\omega(s) = 0$ over $[-t, t]$. We can thus express the ratio between them as follows:

$$\frac{h_{\|}(\omega, t)}{h_{\perp}(\omega, t)} = (d-1) \left(\frac{\int_0^{\pi} \omega(t \cos\theta) \sin^{d-2}\theta \; d\theta}{\int_0^{\pi} \omega(t \cos\theta) \sin^{d}\theta \; d\theta} - 1 \right) = (d-1) \Big(\frac{I_{d-2}}{I_d} - 1 \Big),$$

where we denoted by $I_d = \int_0^{\pi} \omega(t \cos\theta) \sin^d\theta \; d\theta$. Using integration by parts:

$$\begin{aligned}
I_d = &-\omega(t\cos\theta)\cos\theta \sin^{d-1}\theta \Big|_0^{\pi} + \\
&\int_0^{\pi} \cos\theta \Big[\omega(t\cos\theta)(d-1)\sin^{d-2}\theta\cos\theta - \omega'(t\cos\theta) t \sin^d\theta \Big] \; d\theta \\
&= (d-1) \int_0^{\pi} \omega(t\cos\theta)\cos^2\theta \sin^{d-2}\theta \; d\theta - t \int_0^{\pi} \omega'(t\cos\theta)\cos\theta\sin^d\theta \; d\theta \\
&= (d-1) I_{d-2} - (d-1) I_d - t \int_0^{\pi} \omega'(t\cos\theta)\cos\theta\sin^d\theta \; d\theta.
\end{aligned}$$

Because $\omega(s) = 0$ for $s < 0$, the last term reduces to an integral over $[0, \pi/2]$. The resulting integral is non-negative because sine and cosine are both non-negative over that interval. It is 0 if and only if $\omega' = 0$, or equivalently when ω is constant. We have therefore shown that:

$$I_d \leq (d-1) I_{d-2} - (d-1) I_d \implies (d-1) \Big(\frac{I_{d-2}}{I_d} - 1 \Big) \geq 1 \implies \frac{h_{\|}(\omega, t)}{h_{\perp}(\omega, t)} \geq 1,$$

with equality when ω is constant, as desired. □

9.4.1.2 Learning a Codebook

The results above formalize the intuition that the parallel residual plays a more important role in quantization for MIPS. If we were to plug the formalism above into the objective in Equation (9.2) and optimize it to learn a codebook, we would need to compute $h_{\parallel}$ and $h_{\perp}$ using Theorem 9.1. That would prove cumbersome indeed.

Instead, Guo et al. [2020] show that $\omega(s) = \mathbb{1}_{s \geq \theta}$ results in a more computationally-efficient optimization problem. Letting $\eta(t) = \frac{h_{\parallel}(\omega,t)}{h_{\perp}(\omega,t)}$, they show that $\eta/(d-1)$ concentrates around $\frac{(\theta/t)^2}{1-(\theta/t)^2}$ as d becomes larger. So in high dimensions, one can rewrite the objective function of Equation (9.2) as follows:

$$\sum_{u \in \mathcal{X}} \frac{(\theta/\|u\|)^2}{1-(\theta/\|u\|)^2} \|r_{\parallel}(u,\tilde{u})\|^2 + \|r_{\perp}(u,\tilde{u})\|^2.$$

Guo et al. [2020] present an optimization procedure that is based on Lloyd's iterative algorithm for KMeans, and use it to learn a codebook by minimizing the objective above. Empirically, such a codebook outperforms the one that is learnt by optimizing the reconstruction error.

9.4.1.3 Extensions

The score-aware quantization loss has, since its publication, been extended in two different ways. Zhang et al. [2022] adapted the objective function to an Additive Quantization form. Zhang et al. [2023] updated the weight function $\omega(\cdot)$ so that the importance of a data point can be estimated based on a given set of training queries. Both extensions lead to substantial improvements on benchmark datasets.

References

F. André, A.-M. Kermarrec, and N. Le Scouarnec. Cache locality is not enough: High-performance nearest neighbor search with product quantization fast scan. *Proceedings of the VLDB Endowment*, 9(4):288–299, 12 2015.

F. Andre, A.-M. Kermarrec, and N. Le Scouarnec. Quicker adc: Unlocking the hidden potential of product quantization with simd. *IEEE Transactions on Pattern Analysis and Machine Intelligence*, 43(5):1666–1677, 5 2021.

A. Babenko and V. Lempitsky. The inverted multi-index. In *2012 IEEE Conference on Computer Vision and Pattern Recognition*, pages 3069–3076, 2012.

A. Babenko and V. Lempitsky. Additive quantization for extreme vector compression. In *2014 IEEE Conference on Computer Vision and Pattern Recognition*, pages 931–938, 2014.

T. Chen, L. Li, and Y. Sun. Differentiable product quantization for end-to-end embedding compression. In *Proceedings of the 37th International Conference on Machine Learning*, volume 119 of *Proceedings of Machine Learning Research*, pages 1617–1626, 7 2020.

Y. Chen, T. Guan, and C. Wang. Approximate nearest neighbor search by residual vector quantization. *Sensors*, 10(12):11259–11273, 2010.

T. Ge, K. He, Q. Ke, and J. Sun. Optimized product quantization. *IEEE Transactions on Pattern Analysis and Machine Intelligence*, 36(4):744–755, 2014.

R. Gray and D. Neuhoff. Quantization. *IEEE Transactions on Information Theory*, 44(6):2325–2383, 1998.

R. Guo, S. Kumar, K. Choromanski, and D. Simcha. Quantization based fast inner product search. In *Proceedings of the 19th International Conference on Artificial Intelligence and Statistics*, volume 51 of *Proceedings of Machine Learning Research*, pages 482–490, Cadiz, Spain, 5 2016.

R. Guo, P. Sun, E. Lindgren, Q. Geng, D. Simcha, F. Chern, and S. Kumar. Accelerating large-scale inference with anisotropic vector quantization. In *Proceedings of the 37th International Conference on Machine Learning*, 2020.

Y. K. Jang and N. I. Cho. Self-supervised product quantization for deep unsupervised image retrieval. In *Proceedings of the IEEE/CVF International Conference on Computer Vision*, pages 12085–12094, October 2021.

H. Jégou, M. Douze, and C. Schmid. Product quantization for nearest neighbor search. *IEEE Transactions on Pattern Analysis and Machine Intelligence*, 33(1):117–128, 2011.

J. Johnson, M. Douze, and H. Jégou. Billion-scale similarity search with gpus. *IEEE Transactions on Big Data*, 7(3):535–547, 2021.

Y. Kalantidis and Y. Avrithis. Locally optimized product quantization for approximate nearest neighbor search. In *2014 IEEE Conference on Computer Vision and Pattern Recognition*, pages 2329–2336, 2014.

B. Klein and L. Wolf. End-to-end supervised product quantization for image search and retrieval. In *Proceedings of the IEEE/CVF Conference on Computer Vision and Pattern Recognition*, 6 2019.

A. Krishnan and E. Liberty. Projective clustering product quantization, 2021.

M. Liu, Y. Dai, Y. Bai, and L.-Y. Duan. Deep product quantization module for efficient image retrieval. In *IEEE International Conference on Acoustics, Speech and Signal Processing*, pages 4382–4386, 2020.

S. Liu, H. Lu, and J. Shao. Improved residual vector quantization for high-dimensional approximate nearest neighbor search, 2015.

Z. Lu, D. Lian, J. Zhang, Z. Zhang, C. Feng, H. Wang, and E. Chen. Differentiable optimized product quantization and beyond. In *Proceedings of the ACM Web Conference 2023*, pages 3353–3363, 2023.

Y. Matsui, Y. Uchida, H. Jégou, and S. Satoh. A survey of product quantization. *ITE Transactions on Media Technology and Applications*, 6(1): 2–10, 2018.

L. Niu, Z. Xu, L. Zhao, D. He, J. Ji, X. Yuan, and M. Xue. Residual vector product quantization for approximate nearest neighbor search. *Expert Systems with Applications*, 232(C), 12 2023.

M. Norouzi and D. J. Fleet. Cartesian k-means. In *Proceedings of the 2013 IEEE Conference on Computer Vision and Pattern Recognition*, pages 3017–3024, 2013.

E. C. Ozan, S. Kiranyaz, and M. Gabbouj. Competitive quantization for approximate nearest neighbor search. *IEEE Transactions on Knowledge and Data Engineering*, 28(11):2884–2894, 2016.

P. Sun, R. Guo, and S. Kumar. Automating nearest neighbor search configuration with constrained optimization. In *Proceedings of the 11th International Conference on Learning Representations*, 2023.

X. Wu, R. Guo, A. T. Suresh, S. Kumar, D. N. Holtmann-Rice, D. Simcha, and F. Yu. Multiscale quantization for fast similarity search. In *Advances in Neural Information Processing Systems*, volume 30, 2017.

D. Xu, I. W. Tsang, and Y. Zhang. Online product quantization. *IEEE Transactions on Knowledge and Data Engineering*, 30(11):2185–2198, 2018.

T. Yu, J. Yuan, C. Fang, and H. Jin. Product quantization network for fast image retrieval. In *Proceedings of the European Conference on Computer Vision*, September 2018.

J. Zhang, Q. Liu, D. Lian, Z. Liu, L. Wu, and E. Chen. Anisotropic additive quantization for fast inner product search. *Proceedings of the AAAI Conference on Artificial Intelligence*, 36(4):4354–4362, 6 2022.

J. Zhang, D. Lian, H. Zhang, B. Wang, and E. Chen. Query-aware quantization for maximum inner product search. In *Proceedings of the 37th AAAI Conference on Artificial Intelligence*, 2023.

Chapter 10
Sketching

Abstract Sketching is a probabilistic tool to summarize high-dimensional vectors into low-dimensional vectors, called *sketches*, while approximately preserving properties of interest. For example, we may sketch vectors in the Euclidean space such that their L_2 norm is approximately preserved; or sketch points in an inner product space such that the inner product between any two points is maintained with high probability. This chapter reviews a few *data-oblivious* algorithms, cherry-picked from the vast literature on sketching, that are tailored to sparse vectors in an inner product space.

10.1 Intuition

We learnt about quantization as a form of vector compression in Chapter 9. There, vectors are decomposed into L subspaces, with each subspace mapped to C geometrically-cohesive buckets. By coding each subspace into only C values, we can encode an entire vector in $L \log C$ bits, often dramatically reducing the size of a vector collection, though at the cost of losing information in the process.

The challenge, we also learnt, is that not enough can be said about the effects of L, C, and other parameters involved in the process of quantization, on the reconstruction error. We can certainly intuit the asymptotic behavior of quantization, but that is neither interesting nor insightful. That leaves us no option other than settling on a configuration empirically.

Additionally, learning codebooks can become involved and cumbersome. It involves tuning parameters and running clustering algorithms, whose expected behavior is itself ill-understood when handling improper distance functions. The resulting codebooks too may become obsolete in the event of a distributional shift.

S. Bruch, *Foundations of Vector Retrieval*, https://doi.org/10.1007/978-3-031-55182-6_10

This chapter reviews a different class of compression techniques known as *data-oblivious sketching*. Let us break down this phrase and understand each part better.

The data-oblivious qualifier is rather self-explanatory: We make no assumptions about the input data, and in fact, do not even take advantage of the statistical properties of the data. We are, in other words, completely agnostic and oblivious to our input.

While oblivion may put us at a disadvantage and lead to a larger magnitude of error, it creates two opportunities. First, we can often easily quantify the average qualities of the resulting compressed vectors. Second, by design, the compressed vectors are robust under any data drift. Once a vector collection has been compressed, in other words, we can safely assume that any guarantees we were promised will continue to hold.

Sketching, to continue our unpacking of the concept, is a probabilistic tool to reduce the dimensionality of a vector space while preserving certain properties of interest *with high probability.* In its simplest form, sketching is a function $\phi : \mathbb{R}^d \rightarrow \mathbb{R}^{d_\circ}$, where $d_\circ < d$. If the "property of interest" is the Euclidean distance between any pair of points in a collection $\mathcal{X}$, for instance, then $\phi(\cdot)$ must satisfy the following for random points U and V:

$$\mathbb{P}\left[\Big| \lVert \phi(U) - \phi(V) \rVert_2 - \lVert U - V \rVert_2 \Big| < \epsilon \right] > 1 - \delta,$$

for $\delta, \epsilon \in (0, 1)$.

The output of $\phi(u)$, which we call the *sketch* of vector u, is a good substitute for u itself. If all we care about, as we do in top-k retrieval, is the distance between pairs of points, then we retain the ability to deduce that information with high probability just from the sketches of a collection of vectors. Considering that $d_\circ$ is smaller than d, we not only compress the collection through sketching, but, as with quantization, we are able to perform distance computations directly on the compressed vectors.

The literature on sketching offers numerous algorithms that are designed to approximate a wide array of norms, distances, and other properties of data. We refer the reader to the excellent monograph by Woodruff [2014] for a tour of this rich area of research. But to give the reader a better understanding of the connection between sketching and top-k retrieval, we use the remainder of this chapter to delve into three algorithms. To make things more interesting, we specifically review these algorithms in the context of inner product for sparse vectors.

The first is the quintessential linear algorithm due to Johnson and Lindenstrauss [1984]. It is linear in the sense that ϕ is simply a linear transformation, so that $\phi(u) = \Phi u$ for some (random) matrix $\Phi \in \mathbb{R}^{d_\circ \times d}$. We will learn how to construct the required matrix and discuss what guarantees it has to offer.

We then move to two sketching algorithms [Bruch et al., 2023] and [Daliri et al., 2023] whose output space is *not* Euclidean. Instead, the sketch of a vector is a data structure, equipped with a distance function that approximates the inner product between vectors in the original space.

10.2 Linear Sketching with the JL Transform

Let us begin by repeating the well-known result due to Johnson and Lindenstrauss [1984], which we refer to as the JL Lemma:

Lemma 10.1 *For $\epsilon \in (0, 1)$ and any set $\mathcal{X}$ of m points in $\mathbb{R}^d$, and an integer $d_\circ = \Omega(\epsilon^{-2} \ln m)$, there exists a Lipschitz mapping $\phi : \mathbb{R}^d \rightarrow \mathbb{R}^{d_\circ}$ such that*

$$(1 - \epsilon)\|u - v\|_2^2 \leq \|\phi(u) - \phi(v)\|_2^2 \leq (1 + \epsilon)\|u - v\|_2^2,$$

for all $u, v \in \mathcal{X}$.

This result has been studied extensively and further developed since its introduction. Using simple proofs, for example, it can be shown that the mapping ϕ may be a linear transformation by a $d_\circ \times d$ random matrix Φ drawn from a particular class of distributions. Such a matrix Φ is said to form a JL transform.

Definition 10.1 A random matrix $\Phi \in \mathbb{R}^{d_\circ \times d}$ forms a Johnson-Lindenstrauss transform with parameters (ϵ, δ, m), if with probability at least $1 - \delta$, for any m-element subset $\mathcal{X} \subset \mathbb{R}^d$, for all $u, v \in \mathcal{X}$ it holds that $|\langle \Phi u, \Phi v \rangle - \langle u, v \rangle| \leq \epsilon \|u\|_2 \|v\|_2$.

There are many constructions of Φ that form a JL transform. It is trivial to show that when the entries of Φ are independently drawn from $\mathcal{N}(0, \frac{1}{d_\circ})$, then Φ is a JL transform with parameters (ϵ, δ, m) if $d_\circ = \Omega(\epsilon^{-2} \ln(m/\delta))$. In yet another construction, $\Phi = \frac{1}{\sqrt{d_\circ}} R$, where $R \in \{\pm 1\}^{d_\circ \times d}$ is a matrix whose entries are independent Rademacher random variables.

We take the latter as an example due to its simplicity and analyze its properties. As before, we refer the reader to [Woodruff, 2014] for a far more detailed discussion of other (more efficient) constructions of the JL transform.

10.2.1 Theoretical Analysis

We are interested in analyzing the transformation above in the context of inner product. Specifically, we wish to understand what we should expect if, instead of computing the inner product between two vectors u and v in $\mathbb{R}^d$, we perform the operation $\langle Ru, Rv\rangle$ in the transformed space in $\mathbb{R}^{d_\circ}$. Is the outcome an unbiased estimate of the true inner product? How far off may this estimate be? The following result is a first step to answering these questions for two *fixed* vectors.

Theorem 10.1 *Fix two vectors u and $v \in \mathbb{R}^d$. Define $Z_{\text{SKETCH}} = \langle \phi(u), \phi(v)\rangle$ as the random variable representing the inner product of sketches of size $d_\circ$, prepared using the projection $\phi(u) = Ru$, with $R \in \{\pm 1/\sqrt{d_\circ}\}^{d_\circ \times d}$ being a random Rademacher matrix. Z_{SKETCH} is an unbiased estimator of $\langle u, v\rangle$. Its distribution tends to a Gaussian with variance:*

$$\frac{1}{d_\circ}\Big(\|u\|_2^2\|v\|_2^2 + \langle u, v\rangle^2 - 2\sum_i u_i^2 v_i^2\Big).$$

Proof. Consider the random variable $Z = \big(\sum_j R_j u_j\big)\big(\sum_k R_k v_k\big)$, where R_i's are Rademacher random variables. It is clear that $d_\circ Z$ is the product of the sketch coordinate i (for any i): $\phi(u)_i \phi(v)_i$.

We can expand the expected value of Z as follows:

$$\begin{aligned}
\mathbb{E}[Z] &= \mathbb{E}\Big[\big(\sum_j R_j u_j\big)\big(\sum_k R_k v_k\big)\Big] \\
&= \mathbb{E}[\sum_i R_i^2 u_i v_i] + \mathbb{E}[\sum_{j \neq k} R_j R_k u_j v_k] \\
&= \sum_i u_i v_i \underbrace{\mathbb{E}[R_i^2]}_{1} + \sum_{j\neq k} u_j v_k \underbrace{\mathbb{E}[R_j R_k]}_{0} \\
&= \langle u, v\rangle.
\end{aligned}$$

The variance of Z can be expressed as follows:

$$\text{Var}[Z] = \mathbb{E}[Z^2] - \mathbb{E}[Z]^2 = \mathbb{E}\Big[\big(\sum_j R_j u_j\big)^2\big(\sum_k R_k v_k\big)^2\Big] - \langle u, v\rangle^2.$$

We have the following:

$$\mathbb{E}\Big[\big(\sum_j R_j u_j\big)^2\big(\sum_k R_k v_k\big)^2\Big]$$
$$= \mathbb{E}\left[\big(\sum_i u_i^2 + \sum_{i\neq j} R_i R_j u_i u_j\big)\big(\sum_k v_k^2 + \sum_{k\neq l} R_k R_l v_k v_l\big)\right]$$
$$= \|u\|_2^2\|v\|_2^2 + \underbrace{\mathbb{E}\left[\sum_i u_i^2 \sum_{k\neq l} R_k R_l v_k v_l\right]}_{0}$$
$$+ \underbrace{\mathbb{E}\left[\sum_k v_k^2 \sum_{i\neq j} R_i R_j u_i u_j\right]}_{0} + \mathbb{E}\left[\sum_{i\neq j} R_i R_j u_i u_j \sum_{k\neq l} R_k R_l v_k v_l\right]. \qquad (10.1)$$

The last term can be decomposed as follows:

$$\mathbb{E}\Big[\sum_{i\neq j\neq k\neq l} R_i R_j R_k R_l u_i u_j v_k v_l\Big]$$
$$+ \mathbb{E}\Big[\sum_{i=k, j\neq l \lor i\neq k, j=l} R_i R_j R_k R_l u_i u_j v_k v_l\Big]$$
$$+ \mathbb{E}\Big[\sum_{i\neq j, i=k, j=l \lor i\neq j, i=l, j=k} R_i R_j R_k R_l u_i u_j v_k v_l\Big].$$

The first two terms are 0 and the last term can be rewritten as follows:

$$2\,\mathbb{E}\left[\sum_i u_i v_i \big(\sum_j u_j v_j - u_i v_i\big)\right] = 2\langle u, v\rangle^2 - 2\sum_i u_i^2 v_i^2. \qquad (10.2)$$

We now substitute the last term in Equation (10.1) with Equation (10.2) to obtain:

$$\mathrm{Var}[Z] = \|u\|_2^2\|v\|_2^2 + \langle u, v\rangle^2 - 2\sum_i u_i^2 v_i^2.$$

Observe that $Z_{\text{SKETCH}} = 1/d_\circ \sum_i \phi(u)_i \phi(v)_i$ is the sum of independent, identically distributed random variables. Furthermore, for bounded vectors u and v, the variance is finite. By the application of the Central Limit Theorem, we can deduce that the distribution of Z_{SKETCH} tends to a Gaussian distribution with the stated expected value. Noting that $\mathrm{Var}[Z_{\text{SKETCH}}] = 1/d_\circ^2 \sum_i \mathrm{Var}[Z]$ gives the desired result. □

Theorem 10.1 gives a clear model of the inner product error when two fixed vectors are transformed using our particular choice of the JL transform. We learnt that inner product of sketches is an ubiased estimator of the inner product between vectors, and have shown that the error follows a Gaussian distribution.

Let us now position this result in the context of top-k retrieval where the query point is fixed, but the data points are random. To make the analysis more interesting, let us consider sparse vectors, where each coordinate may be 0 with a non-zero probability.

Theorem 10.2 *Fix a query vector $q \in \mathbb{R}^d$ and let X be a random vector drawn according to the following probabilistic model. Coordinate i, X_i, is non-zero with probability $p_i > 0$ and, if it is non-zero, draws its value from a distribution with mean μ and variance $\sigma^2 < \infty$. Then, $Z_{\text{SKETCH}} = \langle \phi(q), \phi(X) \rangle$, with $\phi(u) = Ru$ and $R \in \{\pm 1/\sqrt{d_\circ}\}^{d_\circ \times d}$, has expected value $\mu \sum_i p_i q_i$ and variance:*

$$\frac{1}{d_\circ}\Bigg[(\mu^2+\sigma^2)\Big(\|q\|_2^2\sum_i p_i - \sum_i p_i q_i^2\Big) + \mu^2\Big(\big(\sum_i q_i p_i\big)^2 - \sum_i \big(q_i p_i\big)^2\Big)\Bigg].$$

Proof. It is easy to see that:

$$\mathbb{E}[Z_{\text{SKETCH}}] = \sum_i q_i\, \mathbb{E}[X_i] = \mu \sum_i p_i q_i.$$

As for the variance, we start from Theorem 10.1 and arrive at the following expression:

$$\frac{1}{d_\circ}\Big(\|q\|_2^2\,\mathbb{E}[\|X\|_2^2] + \mathbb{E}[\langle q, X\rangle^2] - 2\sum_i q_i^2\,\mathbb{E}[X_i^2]\Big), \tag{10.3}$$

where the expectation is with respect to X. Let us consider the terms inside the parentheses one by one. The first term becomes:

$$\begin{aligned}\|q\|_2^2\,\mathbb{E}[\|X\|_2^2] &= \|q\|_2^2 \sum_i \mathbb{E}[X_i^2] \\ &= \|q\|_2^2(\mu^2+\sigma^2)\sum_i p_i.\end{aligned}$$

The second term reduces to:

$$\begin{aligned}\mathbb{E}\left[\langle q, X\rangle^2\right] &= \mathbb{E}\left[\langle q, X\rangle\right]^2 + \mathrm{Var}\left[\langle q, X\rangle\right] + \\ &= \mu^2\big(\sum_i q_i p_i\big)^2 + \sum q_i^2\left[(\mu^2+\sigma^2)p_i - \mu^2 p_i^2\right] \\ &= \mu^2\big(\big(\sum_i q_i p_i\big)^2 - \sum_i q_i^2 p_i^2\big) + \sum_i q_i^2 p_i(\mu^2+\sigma^2).\end{aligned}$$

Finally, the last term breaks down to:

$$-2\sum_i q_i^2\,\mathbb{E}[X_i^2] = -2\sum_i q_i^2(\mu^2+\sigma^2)p_i$$
$$= -2(\mu^2+\sigma^2)\sum_i q_i^2 p_i.$$

Putting all these terms back into Equation (10.3) yields the desired expression for variance. □

Let us consider a special case to better grasp the implications of Theorem 10.2. Suppose $p_i = \psi/d$ for some constant ψ for all dimensions i. Further assume, without loss of generality, that the (fixed) query vector has unit norm: $\|q\|_2 = 1$. We can observe that the variance of Z_{SKETCH} decomposes into a term that is $(\mu^2+\sigma^2)(1-1/d)\psi/d_\circ$, and a second term that is a function of $1/d^2$. The mean, on the other hand, is a linear function of the non-zero coordinates in the query: $(\mu\sum_i q_i)\psi/d$. As d grows, the mean of Z_{SKETCH} tends to 0 at a rate proportional to the sparsity rate (ψ/d), while its variance tends to $(\mu^2+\sigma^2)\psi/d_\circ$.

The above suggests that the ability of $\phi(\cdot)$ to preserve the inner product of a query point with a randomly drawn data point deteriorates as a function of the number of non-zero coordinates. For example, when the number of non-zero coordinates becomes larger, $\langle\phi(q),\phi(X)\rangle$ for a fixed query q and a random point X becomes less reliable because the variance of the approximation increases.

10.3 Asymmetric Sketching

Our second sketching algorithm is due to Bruch et al. [2023]. It is unusual in several ways. First, it is designed specifically for retrieval. That is, the objective of the sketching technique is not to preserve the inner product between points in a collection; in fact, as we will learn shortly, the sketch is not even an unbiased estimator. Instead, it is assumed that the setup is retrieval, where we receive a query and wish to *rank* data points in response.

That brings us to its second unusual property: asymmetry. That means, only the data points are sketched while queries remain in the original space. With the help of an asymmetric distance function, however, we can easily compute an *upper-bound* on the query-data point inner product, using the raw query point and the sketch of a data point.

Finally, in its original construction as presented in [Bruch et al., 2023], the sketch was tailored specifically to sparse vectors. As we will show, however, it is trivial to modify the algorithm and adapt it to dense vectors.

In the rest of this section, we will first describe the sketching algorithm for sparse vectors, as well as its extension to dense vectors. We then describe how the distance between a query point in the original space and the sketch of any

Algorithm 5: Sketching of sparse vectors

Input: Sparse vector $u \in \mathbb{R}^d$.
Requirements: h independent random mappings $\pi_o : [d] \to [d_\circ/2]$.
Result: Sketch of u, $\{nz(u);\ \underline{u};\ \overline{u}\}$ consisting of the index of non-zero coordinates of u, the lower-bound sketch, and the upper-bound sketch.

Let $\overline{u}, \underline{u} \in \mathbb{R}^{d_\circ/2}$ be zero vectors
for all $k \in [\frac{d_\circ}{2}]$ **do**
  $\mathcal{I} \leftarrow \{i \in nz(u) \mid \exists\ o\ s.t.\ \pi_o(i) = k\}$
  $\overline{u}_k \leftarrow \max_{i \in \mathcal{I}} u_i$
  $\underline{u}_k \leftarrow \min_{i \in \mathcal{I}} u_i$
end for
return $\{nz(u),\ \underline{u},\ \overline{u}\}$

data point can be computed asymmetrically. Lastly, we review an analysis of the sketching algorithm.

10.3.1 The Sketching Algorithm

Algorithm 5 shows the logic behind the sketching of sparse vectors. It is assumed throughout that the sketch size, $d_\circ$, is even, so that $d_\circ/2$ is an integer. The algorithm also makes use of h independent random mappings $\pi_o : [d] \to [d_\circ/2]$, where each $\pi_o(\cdot)$ projects coordinates in the original space to an integer in the set $[d_\circ/2]$ uniformly randomly.

Intuitively, the sketch of $u \in \mathbb{R}^d$ is a data structure comprising of the index of its set of non-zero coordinates (i.e., $nz(u)$), along with an *upper-bound sketch* ($\overline{u} \in \mathbb{R}^{d_\circ/2}$) and a *lower-bound sketch* ($\underline{u} \in \mathbb{R}^{d_\circ/2}$) on the non-zero values of u. More precisely, the k-th coordinate of $\overline{u}$ ($\underline{u}$) records the largest (smallest) value from the set of all non-zero coordinates in u that map into k according to at least one $\pi_o(\cdot)$.

This sketching algorithm offers a great deal of flexibility. When data vectors are non-negative, we may drop the lower-bounds from the sketch, so that the sketch of u consists only of $\{nz(u), \overline{u}\}$. When vectors are dense, the sketch clearly does not need to store the set of non-zero coordinates, so that the sketch of u becomes $\{\overline{u}, \underline{u}\}$. Finally, when vectors are dense and non-negative, the sketch of u simplifies to $\overline{u}$.

Algorithm 6: Asymmetric distance computation for sparse vectors

Input: Sparse query vector $q \in \mathbb{R}^d$; sketch of data point u: $\{nz(u), \overline{u}, \underline{u}\}$
Requirements: h independent random mappings $\pi_o : [d] \rightarrow [d_\circ/2]$.
Result: Upper-bound on $\langle q, u \rangle$.

$s \leftarrow 0$
for $i \in nz(q) \cap nz(u)$ **do**
  $\mathcal{J} \leftarrow \{\pi_o(i) \mid o \in [h]\}$
  if $q_i > 0$ **then**
    $s \leftarrow s + \min_{j \in \mathcal{J}} \overline{u}_j$
  else
    $s \leftarrow s + \max_{j \in \mathcal{J}} \underline{u}_j$
  end if
end for
return s

10.3.2 Inner Product Approximation

Suppose that we are given a query point $q \in \mathbb{R}^d$ and wish to obtain an estimate of the inner product $\langle q, u \rangle$ for some data vector u. We must do so using only the sketch of u as produced by Algorithm 5. Because the query point is not sketched and, instead, remains in the original d-dimensional space, while u is only known in its sketched form, we say this computation is *asymmetric*. This is not unlike the distance computation between a query point and a quantized data point, as seen in Chapter 9.

This asymmetric procedure is described in Algorithm 6. The algorithm iterates over the intersection of the non-zero coordinates of the query vector and the non-zero coordinates of the data point (which is included in the sketch). It goes without saying that, if the vectors are dense, we may simply iterate over all coordinates. When visiting the i-th coordinate, we first form the set of coordinates that i maps to according to the hash functions π_o's; that is the set $\mathcal{J}$ in the algorithm.

The next step then depends on the sign of the query at that coordinate. When q_i is positive, we find the *least upper-bound* on the value of u_i from its upper-bound sketch. That can be determined by looking at $\overline{u}_j$ for all $j \in \mathcal{J}$, and taking the minimum value among those sketch coordinates. When $q_i < 0$, on the other hand, we find the *greatest lower-bound* instead. In this way, it is always guaranteed that the partial inner product is an upper-bound on the actual partial inner product, $q_i u_i$, as stated in the next theorem.

Theorem 10.3 *The quantity returned by Algorithm 6 is an upper-bound on the inner product of query and data vectors.*

10.3.3 Theoretical Analysis

Theorem 10.3 implies that Algorithm 6 always overestimates the inner product between query and data points. In other words, the inner product approximation error is non-negative. But what can be said about the probability that such an error occurs? How large is the overestimation error? We turn to these questions next.

Before we do so, however, we must agree on a probabilistic model of the data. We follow [Bruch et al., 2023] and assume that a random sparse vector X is drawn from the following distribution. All coordinates of X are mutually independent. Its i-th coordinate is *inactive* (i.e., zero) with probability $1-p_i$. Otherwise, it is *active* and its value is a random variable, X_i, drawn *iid* from some distribution with probability density function (PDF) ϕ and cumulative distribution function (CDF) Φ.

10.3.3.1 Probability of Error

Let us focus on the approximation error of a single active coordinate. Concretely, suppose we have a random vector X whose i-th coordinate is active: $i \in nz(X)$. We are interested in quantifying the likelihood that, if we estimated the value of X_i from the sketch, the estimated value, $\tilde{X}_i$, overshoots or undershoots the actual value.

Formally, we wish to model $\mathbb{P}[\tilde{X}_i \neq X_i]$, Note that, depending on the sign of the query's i-th coordinate $\tilde{X}_i$ may be estimated from the upper-bound sketch ($\overline{X}$), resulting in *overestimation*, or the lower-bound sketch ($\underline{X}$), resulting in *underestimation*. Because the two cases are symmetric, we state the main result for the former case: When $\tilde{X}_i$ is the least upper-bound on X_i, estimated from $\overline{X}$:

$$\tilde{X}_i = \min_{j \in \{\pi_o(i) \mid o \in [h]\}} \overline{X}_j. \tag{10.4}$$

Theorem 10.4 *For large values of $d_\circ$, an active X_i, and $\tilde{X}_i$ estimated using Equation (10.4),*

$$\mathbb{P}\left[\tilde{X}_i > X_i\right] \approx \int \left[1 - \exp\Big(-\frac{2h}{d_\circ}(1-\Phi(\alpha))\sum_{j\neq i} p_j\Big)\right]^h \phi(\alpha)d\alpha,$$

where $\phi(\cdot)$ and $\Phi(\cdot)$ are the PDF and CDF of X_i.

Extending this result to the lower-bound sketch involves replacing $1-\Phi(\alpha)$ with $\Phi(\alpha)$. When the distribution defined by ϕ is symmetric, the probabilities of error too are symmetric for the upper-bound and lower-bound sketches.

Proof of Theorem 10.4. Recall that $\tilde{X}_i$ is estimated as follows:

$$\tilde{X}_i = \min_{j \in \{\pi_o(i) \mid o \in [h]\}} \overline{X}_j.$$

So we must look up $\overline{X}_j$ for values of j produced by π_o's.

Suppose one such value is k (i.e., $k = \pi_o(i)$ for some $o \in [h]$). The event that $\overline{X}_k > X_i$ happens only when there exists another active coordinate X_j such that $X_j > X_i$ *and* $\pi_o(j) = k$ for some π_o.

To derive $\mathbb{P}[\overline{X}_k > X_i]$, it is easier to think in terms of complementary events: $\overline{X}_k = X_i$ if every other active coordinate whose value is larger than X_i maps to a sketch coordinate *except* k. Clearly the probability that any arbitrary X_j maps to a sketch coordinate other than k is simply $1 - 2/d_\circ$. Therefore, given a vector X, the probability that no active coordinate X_j larger than X_i maps to the k-th coordinate of the sketch, which we denote by "Event A," is:

$$\mathbb{P}\Big[\text{Event A} \mid X\Big] = 1 - (1 - \frac{2}{d_\circ})^{h \sum_{j \neq i} \mathbb{1}_{X_j \text{ is active}} \mathbb{1}_{X_j > X_i}}.$$

Because $d_\circ$ is large by assumption, we can approximate $e^{-1} \approx (1 - 2/d_\circ)^{d_\circ/2}$ and rewrite the expression above as follows:

$$\mathbb{P}\left[\text{Event A} \mid X\right] \approx 1 - \exp\Big(- \frac{2h}{d_\circ} \sum_{j \neq i} \mathbb{1}_{X_j \text{ is active}} \mathbb{1}_{X_j > X_i}\Big).$$

Finally, we marginalize the expression above over X_j's for $j \neq i$ to remove the dependence on all but the i-th coordinate of X. To simplify the expression, however, we take the expectation over the first-order Taylor expansion of the right hand side around 0. This results in the following approximation:

$$\mathbb{P}\left[\text{Event A} \mid X_i = \alpha\right] \approx 1 - \exp\Big(- \frac{2h}{d_\circ}(1 - \Phi(\alpha)) \sum_{j \neq i} p_j\Big).$$

For $\tilde{X}_i$ to be larger than X_i, event A must take place for all h sketch coordinates. That probability, by the independence of random mappings, is:

$$\mathbb{P}\left[\tilde{X}_i > X_i \mid X_i = \alpha\right] \approx \left[1 - \exp\Big(- \frac{2h}{d_\circ}(1 - \Phi(\alpha)) \sum_{j \neq i} p_j\Big)\right]^h.$$

In deriving the expression above, we conditioned the event on the value of X_i. Taking the marginal probability leads us to the following expression for the event that $\tilde{X}_i > X_i$ for any i, concluding the proof:

$$\begin{aligned}\mathbb{P}\left[\tilde{X}_i > X_i\right] &\approx \int \Big[1 - \exp\Big(-\frac{2h}{d_\circ}(1-\Phi(\alpha))\sum_{j\neq i} p_j\Big)\Big]^h d\,\mathbb{P}(\alpha) \\ &\approx \int \Big[1 - \exp\Big(-\frac{2h}{d_\circ}(1-\Phi(\alpha))\sum_{j\neq i} p_j\Big)\Big]^h \phi(\alpha)d\alpha.\end{aligned}$$

□

Theorem 10.4 offers insights into the behavior of the upper-bound sketch. The first observation is that the sketching mechanism presented here is more suitable for distributions where larger values occur with a smaller probability such as sub-Gaussian variables. In such cases, the larger the value is, the smaller its chance of being overestimated by the upper-bound sketch. Regardless of the underlying distribution, empirically, the largest value in a vector is always estimated *exactly*.

The second insight is that there is a sweet spot for h given a particular value of $d_\circ$: using more random mappings helps lower the probability of error until the sketch starts to saturate, at which point the error rate increases. This particular property is similar to the behavior of a Bloom filter [Bloom, 1970].

10.3.3.2 Distribution of Error

We have modeled the probability that the sketch of a vector overestimates a value. In this section, we examine the shape of the distribution of error in the form of its CDF. Formally, assuming X_i is active and $\tilde{X}_i$ is estimated using Equation (10.4), we wish to find an expression for $\mathbb{P}[|\tilde{X}_i - X_i| < \epsilon]$ for any $\epsilon > 0$.

Theorem 10.5 *Suppose X_i is active and draws its value from a distribution with PDF and CDF ϕ and Φ. Suppose further that $\tilde{X}_i$ is the least upper-bound on X_i, obtained using Equation (10.4). Then:*

$$\mathbb{P}[\tilde{X}_i - X_i \leq \epsilon] \approx 1 - \int \Big[1 - \exp\Big(-\frac{2h}{d_\circ}(1-\Phi(\alpha+\epsilon))\sum_{j\neq i} p_j\Big)\Big]^h \phi(\alpha)d\alpha.$$

Proof. We begin by quantifying the conditional probability $\mathbb{P}[\tilde{X}_i - X_i \leq \epsilon \mid X_i = \alpha]$. Conceptually, the event in question happens when all values that collide with X_i are less than or equal to $X_i + \epsilon$. This event can be characterized as the complement of the event that all h sketch coordinates that contain X_i collide with values greater than $X_i + \epsilon$. Using this complementary event, we can write the conditional probability as follows:

$$\mathbb{P}[\tilde{X}_i - X_i \leq \epsilon \mid X_i = \alpha] = 1 - \big[1 - (1 - \frac{2}{d_\circ})^{h(1-\Phi(\alpha+\epsilon))\sum_{j\neq i} p_j}\big]^h$$

$$\approx 1 - \Bigg[1 - \exp\Big(-\frac{2h}{d_\circ}(1-\Phi(\alpha+\epsilon))\sum_{j\neq i} p_j\Big)\Bigg]^h.$$

We complete the proof by computing the marginal distribution over the support. □

Given the CDF of $\tilde{X}_i - X_i$ and the fact that $\tilde{X}_i - X_i \geq 0$, it follows that its expected value conditioned on X_i being active is:

Lemma 10.2 *Under the conditions of Theorem 10.5:*

$$\mathbb{E}[\tilde{X}_i - X_i] \approx \int_0^\infty \int \Bigg[1 - \exp\Big(-\frac{2h}{d_\circ}(1-\Phi(\alpha+\epsilon))\sum_{j\neq i} p_j\Big)\Bigg]^h \phi(\alpha)\, d\alpha\, d\epsilon.$$

10.3.3.3 Case Study: Gaussian Vectors

Let us make the analysis more concrete by applying the results to random Gaussian vectors. In other words, suppose all active X_i's are drawn from a zero-mean, unit-variance Gaussian distribution. We can derive a closed-form expression for the overestimation probability as the following corollary shows.

Corollary 10.1 *Suppose the probability that a coordinate is active, p_i, is equal to p for all coordinates of the random vector $X \in \mathbb{R}^d$. When an active X_i, drawn from $\mathcal{N}(0,1)$, is estimated using the upper-bound sketch with Equation (10.4), the overestimation probability is:*

$$\mathbb{P}\big[\tilde{X}_i > X_i\big] \approx 1 + \sum_{k=1}^{h} \binom{h}{k}(-1)^k \frac{d_\circ}{2kh(d-1)p}\big(1 - e^{-\frac{2kh(d-1)p}{d_\circ}}\big).$$

We begin by proving the special case where $h = 1$.

Lemma 10.3 *Under the conditions of Corollary 10.1 with $h = 1$, the probability that the upper-bound sketch overestimates the value of X_i is:*

$$\mathbb{P}\big[\tilde{X}_i > X_i\big] \approx 1 - \frac{d_\circ}{2(d-1)p}\big(1 - e^{-\frac{2(d-1)p}{d_\circ}}\big).$$

Proof. From Theorem 10.4 we have that:

$$\mathbb{P}\big[\tilde{X}_i > X_i\big] \approx \int \big[1 - e^{-\frac{2h}{d_\circ}(1-\Phi(\alpha))(d-1)p}\big]^h d\,\mathbb{P}(\alpha)$$

$$\overset{h=1}{=} \int \big[1 - e^{-\frac{2(1-\Phi(\alpha))(d-1)p}{d_\circ}}\big] d\,\mathbb{P}(\alpha).$$

Given that X_i's are drawn from a Gaussian distribution, and using the approximation above, we can rewrite the probability of error as:

$$\mathbb{P}\left[\tilde{X}_i > X_i\right] \approx \frac{1}{\sqrt{2\pi}} \int_{-\infty}^{\infty} \left[1 - e^{-\frac{2(d-1)p}{d_\circ}(1-\Phi(\alpha))}\right] e^{-\frac{\alpha^2}{2}} d\alpha.$$

We now break up the right hand side into the following three sums, replacing $2(d-1)p/d_\circ$ with β for brevity:

$$\mathbb{P}\left[\tilde{X}_i > X_i\right] \approx \int_{-\infty}^{\infty} \frac{1}{\sqrt{2\pi}} e^{-\frac{\alpha^2}{2}} d\alpha \tag{10.5}$$

$$- \int_{-\infty}^{0} \frac{1}{\sqrt{2\pi}} e^{-\beta(1-\Phi(\alpha))} e^{-\frac{\alpha^2}{2}} d\alpha \tag{10.6}$$

$$- \int_{0}^{\infty} \frac{1}{\sqrt{2\pi}} e^{-\beta(1-\Phi(\alpha))} e^{-\frac{\alpha^2}{2}} d\alpha. \tag{10.7}$$

The sum in (10.5) is equal to the quantity 1. Let us turn to (10.7) first. We have that:

$$1 - \Phi(\alpha) \overset{\alpha \geq 0}{=} \frac{1}{2} - \underbrace{\int_0^{\alpha} \frac{1}{\sqrt{2\pi}} e^{-\frac{t^2}{2}} dt}_{\lambda(\alpha)}.$$

As a result, we can write:

$$\begin{aligned}
\int_0^{\infty} \frac{1}{\sqrt{2\pi}} e^{-\beta(1-\Phi(\alpha))} e^{-\frac{\alpha^2}{2}} d\alpha &= \int_0^{\infty} \frac{1}{\sqrt{2\pi}} e^{-\beta(\frac{1}{2}-\lambda(\alpha))} e^{-\frac{\alpha^2}{2}} d\alpha \\
&= e^{-\frac{\beta}{2}} \int_0^{\infty} \frac{1}{\sqrt{2\pi}} e^{\beta\lambda(\alpha)} e^{-\frac{\alpha^2}{2}} d\alpha \\
&= \frac{1}{\beta} e^{-\frac{\beta}{2}} e^{\beta\lambda(\alpha)} \Big|_0^{\infty} = \frac{1}{\beta} e^{-\frac{\beta}{2}} \left(e^{\frac{\beta}{2}} - 1\right) \\
&= \frac{1}{\beta} \left(1 - e^{-\frac{\beta}{2}}\right)
\end{aligned}$$

By similar reasoning, and noting that:

$$1 - \Phi(\alpha) \overset{\alpha \leq 0}{=} \frac{1}{2} + \underbrace{\int_{\alpha}^{0} \frac{1}{\sqrt{2\pi}} e^{-\frac{t^2}{2}} dt}_{-\lambda(\alpha)},$$

we arrive at:

$$\int_{-\infty}^{0} \frac{1}{\sqrt{2\pi}} e^{-\beta(1-\Phi(\alpha))} e^{-\frac{\alpha^2}{2}} d\alpha = \frac{1}{\beta} e^{-\frac{\beta}{2}} \left(1 - e^{-\frac{\beta}{2}}\right)$$

Plugging the results above into Equations (10.5), (10.6), and (10.7) results in:

$$\begin{aligned}\mathbb{P}\left[\tilde{X}_i > X_i\right] &\approx 1 - \frac{1}{\beta}\big(1 - e^{-\frac{\beta}{2}}\big) - \frac{1}{\beta}e^{-\frac{\beta}{2}}\big(1 - e^{-\frac{\beta}{2}}\big) \\ &= 1 - \frac{1}{\beta}\big(1 - e^{-\frac{\beta}{2}}\big)\big(1 + e^{-\frac{\beta}{2}}\big) \\ &= 1 - \frac{d_\circ}{2(d-1)p}\big(1 - e^{-\frac{2(d-1)p}{d_\circ}}\big),\end{aligned}$$

which completes the proof. □

Given the result above, the solution for the general case of $h > 0$ is straightforward to obtain.

Proof of Corollary 10.1. Using the binomial theorem, we have that:

$$\begin{aligned}\mathbb{P}\left[\tilde{X}_i > X_i\right] &\approx \int \left[1 - e^{-\frac{2h}{d_\circ}(1-\Phi(\alpha))(d-1)p}\right]^h d\,\mathbb{P}(\alpha) \\ &= \sum_{k=0}^{h} \binom{h}{k} \int \big(-e^{-\frac{2h}{d_\circ}(1-\Phi(\alpha))(d-1)p}\big)^k d\,\mathbb{P}(\alpha).\end{aligned}$$

We rewrite the expression above for Gaussian variables to arrive at:

$$\mathbb{P}\left[\tilde{X}_i > X_i\right] \approx \frac{1}{\sqrt{2\pi}} \sum_{k=0}^{h} \binom{h}{k} \int_{-\infty}^{\infty} \big(-e^{-\frac{2h(d-1)p}{d_\circ}(1-\Phi(\alpha))}\big)^k e^{-\frac{\alpha^2}{2}}\,d\alpha.$$

Following the proof of the previous lemma, we can expand the right hand side as follows:

$$\begin{aligned}\mathbb{P}\left[\tilde{X}_i > X_i\right] &\approx 1 + \frac{1}{\sqrt{2\pi}} \sum_{k=1}^{h} \binom{h}{k} (-1)^k \int_{-\infty}^{\infty} e^{-\frac{2kh(d-1)p}{d_\circ}(1-\Phi(\alpha))} e^{-\frac{\alpha^2}{2}}\,d\alpha \\ &= 1 + \sum_{k=1}^{h} \binom{h}{k} (-1)^k \frac{d_\circ}{2kh(d-1)p}\big(1 - e^{-\frac{2kh(d-1)p}{d_\circ}}\big),\end{aligned}$$

which completes the proof. □

Let us now consider the CDF of the overestimation error.

Corollary 10.2 *Under the conditions of Corollary 10.1 the CDF of overestimation error for an active coordinate $X_i \sim \mathcal{N}(0, \sigma)$ is:*

$$\mathbb{P}[\tilde{X}_i - X_i \leq \epsilon] \approx 1 - \left[1 - \exp\Big(-\frac{2h(d-1)p}{d_\circ}(1 - \Phi'(\epsilon))\Big)\right]^h,$$

where $\Phi'(\cdot)$ is the CDF of a zero-mean Gaussian with standard deviation $\sigma\sqrt{2}$.

Proof. When the active values of a vector are drawn from a Gaussian distribution, then the pairwise difference between any two coordinates has a Gaussian distribution with standard deviation $\sqrt{\sigma^2 + \sigma^2} = \sigma\sqrt{2}$. As such, we may estimate $1 - \Phi(\alpha + \epsilon)$ by considering the probability that a pair of coordinates (one of which having value α) has a difference greater than ϵ: $\mathbb{P}[X_i - X_j > \epsilon]$. With that idea, we may thus write:

$$1 - \Phi(\alpha + \epsilon) = 1 - \Phi'(\epsilon).$$

The claim follows by using the above identity in Theorem 10.5. □

Corollary 10.2 enables us to find a particular sketch configuration given a desired bound on the probability of error, as the following lemma shows.

Lemma 10.4 *Under the conditions of Corollary 10.2, and given a choice of* $\epsilon, \delta \in (0, 1)$ *and the number of random mappings* h, $\mathbb{P}[\tilde{X}_i - X_i \leq \epsilon]$ *with probability at least* $1 - \delta$ *if:*

$$d_\circ > -\frac{2h(d-1)p(1 - \Phi'(\epsilon))}{\log(1 - \delta^{1/h})}.$$

10.3.3.4 Error of Inner Product

We have thus far quantified the probability that a value estimated from the upper-bound sketch overestimates the original value of a randomly chosen coordinate. We also characterized the distribution of the overestimation error for a single coordinate and derived expressions for special distributions. In this section, we quantify the overestimation error when approximating the inner product between a fixed query point and a random data point using Algorithm 6.

To make the notation less cluttered, however, let us denote by $\tilde{X}_i$ our estimate of X_i. The estimated quantity is 0 if $i \notin nz(X)$. Otherwise, it is estimated either from the upper-bound sketch or the lower-bound sketch, depending on the sign of q_i. Finally denote by $\tilde{X}$ a reconstruction of X where each $\tilde{X}_i$ is estimated as described above.

Consider the expected value of $\tilde{X}_i - X_i$ conditioned on X_i being active—that is a quantity we analyzed previously. Let $\mu_i = \mathbb{E}[\tilde{X}_i - X_i;\ X_i \text{ is active}]$. Similarly denote by σ_i^2 its variance when X_i is active. Given that X_i is active with probability p_i and inactive with probability $1 - p_i$, it is easy to show that $\mathbb{E}[\tilde{X}_i - X_i] = p_i\mu_i$ (note we have removed the condition on X_i being active) and that its variance $\mathrm{Var}[\tilde{X}_i - X_i] = p_i\sigma_i^2 + p_i(1 - p_i)\mu_i^2$.

With the above in mind, we state the following result.

Theorem 10.6 *Suppose that* $q \in \mathbb{R}^d$ *is a sparse vector. Suppose in a random sparse vector* $X \in \mathbb{R}^d$, *a coordinate* X_i *is active with probability* p_i *and, when active, draws its value from some well-behaved distribution (i.e., with finite* *expectation, variance, and third moment). If* $\mu_i = \mathbb{E}[\tilde{X}_i - X_i;\ X_i \text{ is active}]$ *and* $\sigma_i^2 = \mathrm{Var}[\tilde{X}_i - X_i;\ X_i \text{ is active}]$, *then the random variable* Z *defined as follows:*

$$Z \triangleq \frac{\langle q, \tilde{X} - X \rangle - \sum_{i \in nz(q)} q_i p_i \mu_i}{\sqrt{\sum_{i \in nz(q)} q_i^2 \left(p_i \sigma_i^2 + p_i(1 - p_i)\mu_i^2\right)}}, \tag{10.8}$$

approximately tends to a standard Gaussian distribution as $|nz(q)|$ *grows.*

Proof. Let us expand the inner product between q and $\tilde{X} - X$ as follows:

$$\langle q, \tilde{X} - X \rangle = \sum_{i \in nz(q)} q_i \underbrace{\left(\tilde{X}_i - X_i\right)}_{Z_i}. \tag{10.9}$$

The expected value of $\langle q, \tilde{X} - X \rangle$ is:

$$\mathbb{E}[\langle q, \tilde{X} - X \rangle] = \sum_{i \in nz(q)} q_i \, \mathbb{E}[Z_i] = \sum_{i \in nz(q)} q_i p_i \mu_i.$$

Its variance is:

$$\mathrm{Var}[\langle q, \tilde{X} - X \rangle] = \sum_{i \in nz(q)} q_i^2 \, \mathrm{Var}[Z_i] = \sum_{i \in nz(q)} q_i^2 \left(p_i \sigma_i^2 + p_i(1 - p_i)\mu_i^2\right)$$

Because we assumed that the distribution of X_i is well-behaved, we can conclude that $\mathrm{Var}[Z_i] > 0$ and that $\mathbb{E}[|Z_i|^3] < \infty$. If we operated on the assumption that $q_i Z_i$'s are independent—in reality, they are weakly dependent—albeit not identically distributed, we can appeal to the Berry-Esseen theorem to complete the proof. □

10.3.4 Fixing the Sketch Size

It is often desirable for a sketching algorithm to produce a sketch with a constant size. That makes the size of a collection of sketches predictable, which is often required for resource allocation. Algorithm 5, however, produces a sketch whose size is variable. That is because the sketch contains the set of non-zero coordinates of the vector.

It is, however, straightforward to fix the sketch size. The key to that is the fact that Algorithm 6 uses $nz(u)$ of a vector u only to ascertain if a query's non-zero coordinates are present in the vector u. In effect, all the sketch must provide is a mechanism to perform set membership tests. That is precisely what fixed-size signatures such as Bloom filters [Bloom, 1970] do, albeit probabilistically.

Algorithm 7: Sketching with threshold sampling

Input: Vector $u \in \mathbb{R}^d$.
Requirements: a random mapping $\pi : [d] \to [0, 1]$.
Result: Sketch of u, $\{\mathcal{I}, \mathcal{V}, \|u\|_2^2\}$ consisting of the index and value of sampled coordinates in $\mathcal{I}$ and $\mathcal{V}$, and the squared norm of the vector.

$\mathcal{I}, \mathcal{V} \leftarrow \emptyset$
for $i \in nz(u)$ **do**
  $\theta \leftarrow d_\circ \frac{u_i^2}{\|u\|_2^2}$
  if $\pi(i) \leq \theta$ **then**
    Append i to $\mathcal{I}$, u_i to $\mathcal{V}$
  end if
end for
return $\{\mathcal{I}, \mathcal{V}, \|u\|_2^2\}$

10.4 Sketching by Sampling

Our final sketching algorithm is designed specifically for inner product and is due to Daliri et al. [2023]. The guiding principle is simple: coordinates with larger values contribute more heavily to inner product than coordinates with smaller values. That is an obvious fact that is a direct result of the linearity of inner product: $\langle u, v \rangle = \sum_i u_i v_i$.

Daliri et al. [2023] use that insight as follows. When forming the sketch of vector u, they sample coordinates (without replacement) from u according to a distribution defined by the magnitude of each coordinate. Larger values are given a higher chance of being sampled, while smaller values are less likely to be selected. The sketch, in the end, is a data structure that is made up of the index of sampled coordinates, their values, and additional statistics.

The research question here concerns the sampling process: How must we sample coordinates such that any distance computed from the sketch is an unbiased estimate of the inner product itself? The answer to that question also depends, of course, on how we compute the distance from a pair of sketches. Considering the non-linearity of the sketch, distance computation can no longer be the inner product of sketches.

In the remainder of this section, we review the sketching algorithm, describe distance computation given sketches, and analyze the expected error. In our presentation, we focus on the simpler variant of the algorithm proposed by Daliri et al. [2023], dubbed "threshold sampling."

10.4.1 The Sketching Algorithm

Algorithm 7 presents the "threshold sampling" sketching technique by Daliri et al. [2023]. It is assumed throughout that the desired sketch size is $d_\circ$, and

Algorithm 8: Distance computation for threshold sampling

Input: Sketches of vectors u and v: $\{\mathcal{I}_u, \mathcal{V}_u, \|u\|_2^2\}$ and $\{\mathcal{I}_v, \mathcal{V}_v, \|v\|_2^2\}$.
Result: An unbiased estimate of $\langle u, v\rangle$.

$s \leftarrow 0$
for $i \in \mathcal{I}_u \cap \mathcal{I}_v$ **do**
  $s \leftarrow s + u_i v_i / \min(1, d_\circ u_i^2/\|u\|_2^2, d_\circ v_i^2/\|v\|_2^2)$
end for
return s

that the algorithm has access to a random hash function π that maps integers in $[d]$ to the unit interval.

The algorithm iterates over all non-zero coordinates of the input vector and makes a decision as to whether that coordinate should be added to the sketch. The decision is made based on the relative magnitude of the coordinate, as weighted by $u_i^2/\|u\|_2^2$. If u_i^2 is large, coordinate i has a higher chance of being sampled, as desired.

Notice, however, that the target sketch size $d_\circ$ is realized *in expectation* only. In other words, we may end up with more than $d_\circ$ coordinates in the sketch, or we may have fewer entries. Daliri et al. [2023] propose a different variant of the algorithm that is guaranteed to give a fixed sketch size; we refer the reader to their work for details.

10.4.2 Inner Product Approximation

When sketching a vector using a JL transform, we simply get a vector in the $d_\circ$-dimensional Euclidean space, where inner product is well-defined. So if $\phi(u)$ and $\phi(v)$ are sketches of two d-dimensional vectors u and v, we approximate $\langle u, v\rangle$ with $\langle \phi(u), \phi(v)\rangle$. It could not be more straightforward.

A sketch produced by Algorithm 7, however, is not as nice. Approximating $\langle u, v\rangle$ from their sketches requires a custom distance function defined for the sketch. That is precisely what Algorithm 8 outlines.

In the algorithm, it is understood that u_i and v_i corresponding to $i \in \mathcal{I}_u \cap \mathcal{I}_v$ are present in $\mathcal{V}_u$ and $\mathcal{V}_v$, respectively. These quantities, along with $d_\circ$ and the norms of the vectors are used to weight each partial inner product. The final quantity, as we will learn shortly, is an unbiased estimate of the inner product between u and v.

10.4.3 Theoretical Analysis

Theorem 10.7 *Algorithm 7 produces sketches that consist of at most $d_\circ$ coordinates in expectation.*

Proof. The number of sampled coordinates is $|\mathcal{I}|$. That quantity can be expressed as follows:

$$|\mathcal{I}| = \sum_{i=1}^{d} \mathbb{1}_{i \in \mathcal{I}}.$$

Taking expectation of both sides and using the linearity of expectation, we obtain the following:

$$\mathbb{E}[|\mathcal{I}|] = \sum_i \mathbb{E}[\mathbb{1}_{i \in \mathcal{I}}] = \sum_i \min(1, d_\circ \frac{u_i^2}{\|u\|_2^2}) \leq d_\circ.$$

□

Theorem 10.8 *Algorithm 8 yields an unbiased estimate of inner product.*

Proof. From the proof of the previous theorem, we know that coordinate i of an arbitrary vector u is included in the sketch with probability equal to:

$$\min(1, d_\circ \frac{u_i^2}{\|u\|_2^2}).$$

As such, the odds that $i \in \mathcal{I}_u \cap \mathcal{I}_v$ is:

$$p_i = \min(1, d_\circ \frac{u_i^2}{\|u\|_2^2}, d_\circ \frac{v_i^2}{\|v\|_2^2}).$$

Algorithm 8 gives us a weighted sum of the coordinates that are present in $\mathcal{I}_u \cap \mathcal{I}_v$. We can rewrite that sum using indicator functions as follows:

$$\sum_{i=1}^{d} \mathbb{1}_{i \in \mathcal{I}_u \cap \mathcal{I}_v} \frac{u_i v_i}{p_i}.$$

In expectation, then:

$$\mathbb{E}\Big[\sum_{i=1}^{d} \mathbb{1}_{i \in \mathcal{I}_u \cap \mathcal{I}_v} \frac{u_i v_i}{p_i}\Big] = \sum_{i=1}^{d} p_i \frac{u_i v_i}{p_i} = \langle u, v \rangle,$$

as required. □

Theorem 10.9 *If S is the output of Algorithm 8 for sketches of vectors u and v, then:*

$$\mathrm{Var}[S] \leq \frac{2}{d_\circ} \max\left(\|u_*\|_2^2\|v\|_2^2, \|u\|_2^2\|v_*\|_2^2\right),$$

where u_* and v_* are the vectors u and v restricted to the set of non-zero coordinates common to both vectors (i.e., $* = \{i \mid u_i \neq 0 \wedge v_i \neq 0\}$).

Proof. We use the same proof strategy as in the previous theorem. In particular, we write:

$$\begin{aligned}\mathrm{Var}[S] = \mathrm{Var}\left[\sum_{i\in *} \mathbb{1}_{i\in\mathcal{I}_u\cap\mathcal{I}_v}\frac{u_iv_i}{p_i}\right] &= \sum_{i\in *}\mathrm{Var}\left[\mathbb{1}_{i\in\mathcal{I}_u\cap\mathcal{I}_v}\frac{u_iv_i}{p_i}\right] \\ &= \sum_{i\in *}\frac{u_i^2v_i^2}{p_i^2}\mathrm{Var}\left[\mathbb{1}_{i\in\mathcal{I}_u\cap\mathcal{I}_v}\right].\end{aligned}$$

Turning to the term inside the sum, we obtain:

$$\mathrm{Var}\left[\mathbb{1}_{i\in\mathcal{I}_u\cap\mathcal{I}_v}\right] = p_i - p_i^2,$$

which is 0 if $p_i = 1$ and less than p_i otherwise. Putting everything together, we complete the proof:

$$\begin{aligned}\mathrm{Var}[S] &\leq \sum_{i\in *,\ p_i\neq 1}\frac{u_i^2v_i^2}{p_i} = \|u\|_2^2\|v\|_2^2\sum_{i\in *,\ p_i\neq 1}\frac{\left(u_i^2/\|u\|_2^2\right)\left(v_i^2/\|v\|_2^2\right)}{d_\circ\min\left(u_i^2/\|u\|_2^2, v_i^2/\|v\|_2^2\right)} \\ &= \frac{\|u\|_2^2\|v\|_2^2}{d_\circ}\sum_{i\in *,\ p_i\neq 1}\max\left(u_i^2/\|u\|_2^2, v_i^2/\|v\|_2^2\right) \\ &= \frac{\|u\|_2^2\|v\|_2^2}{d_\circ}\sum_{i\in *}\frac{u_i^2}{\|u\|_2^2}+\frac{v_i^2}{\|v\|_2^2} \\ &= \frac{\|u\|_2^2\|v\|_2^2}{d_\circ}\left(\frac{\|u_*\|_2^2}{\|u\|_2^2}+\frac{\|v_*\|_2^2}{\|v\|_2^2}\right) \\ &= \frac{1}{d_\circ}\left(\|u_*\|_2^2\|v\|_2^2+\|u\|_2^2\|v_*\|_2^2\right) \\ &\leq \frac{2}{d_\circ}\max\left(\|u_*\|_2^2\|v\|_2^2, \|u\|_2^2\|v_*\|_2^2\right).\end{aligned}$$

□

Theorem 10.9 tells us that, if we estimated $\langle u, v\rangle$ for two vectors u and v using Algorithm 8, then the variance of our estimate will be bounded by factors that depend on the non-zero coordinates that u and v have in common. Because $nz(u) \cap nz(v)$ has at most d entries, estimates of inner product based on Threshold Sampling should generally be more accurate than those obtained from JL sketches. This is particularly the case when u and v are sparse.

References

B. H. Bloom. Space/time trade-offs in hash coding with allowable errors. *Commun. ACM*, 13(7):422–426, jul 1970.

S. Bruch, F. M. Nardini, A. Ingber, and E. Liberty. An approximate algorithm for maximum inner product search over streaming sparse vectors. *ACM Transactions on Information Systems*, 42(2), nov 2023.

M. Daliri, J. Freire, C. Musco, A. Santos, and H. Zhang. Sampling methods for inner product sketching, 2023.

W. B. Johnson and J. Lindenstrauss. Extensions of lipschitz mappings into hilbert space. *Contemporary Mathematics*, 26:189–206, 1984.

D. P. Woodruff. Sketching as a tool for numerical linear algebra. *Foundations and Trends in Theoretical Computer Science*, 10(1–2):1–157, Oct 2014. ISSN 1551-305X.

Part IV
Appendices

Appendix A
Collections

Abstract This appendix gives a description of the vector collections used in experiments throughout this monograph. These collections demonstrate different operating points in a typical use-case. For example, some consist of dense vectors, others of sparse vectors; some have few dimensions and others are in much higher dimensions; some are relatively small while others contain a large number of points.

Table A.1 gives a description of the dense vector collections used throughout this monograph and summarizes their key statistics.

Table A.1: Dense collections used in this monograph along with select statistics.

Collection	Vector Count	Query Count	Dimensions
GloVe-25 [Pennington et al., 2014]	1.18M	10,000	25
GloVe-50	1.18M	10,000	50
GloVe-100	1.18M	10,000	100
GloVe-200	1.18M	10,000	200
Deep1b [Yandex and Lempitsky, 2016]	9.99M	10,000	96
MS Turing [Zhang et al., 2019]	10M	100,000	100
Sift [Lowe, 2004]	1M	10,000	128
Gist [Oliva and Torralba, 2001]	1M	1,000	960

In addition to the vector collections above, we convert a few text collections into vectors using various embedding models. These collections are described in Table A.2. Please see [Bajaj et al., 2018] for a complete description of the MS MARCO v1 collection and [Thakur et al., 2021] for the others.

When transforming the text collections of Table A.2 into vectors, we use the following embedding models:

S. Bruch, *Foundations of Vector Retrieval*, https://doi.org/10.1007/978-3-031-55182-6_11

Table A.2: Text collections along with key statistics. The rightmost two columns report the average number of non-zero entries in data points and, in parentheses, queries for sparse vector representations of the collections.

Collection	Vector Count	Query Count	Splade	Efficient Splade
MS Marco Passage	8.8M	6,980	127 (49)	185 (5.9)
NQ	2.68M	3,452	153 (51)	212 (8)
Quora	523K	10,000	68 (65)	68 (8.9)
HotpotQA	5.23M	7,405	131 (59)	125 (13)
Fever	5.42M	6,666	145 (67)	140 (8.6)
DBPedia	4.63M	400	134 (49)	131 (5.9)

- AllMiniLM-l6-v2:[1] Projects text documents into 384-dimensional dense vectors for retrieval with angular distance.
- Tas-B [Hofstätter et al., 2021]: A bi-encoder model that was trained using supervision from a cross-encoder and a ColBERT [Khattab and Zaharia, 2020] model, and produces 768-dimensional dense vectors that are meant for MIPS. The checkpoint used in this work is available on HuggingFace.[2]
- Splade [Formal et al., 2022]:[3] Produces sparse representations for text. The vectors have roughly 30,000 dimensions, where each dimension corresponds to a term in the BERT [Devlin et al., 2019] WordPiece [Wu et al., 2016] vocabulary. Non-zero entries in a vector reflect learnt term importance weights.
- Efficient Splade [Lassance and Clinchant, 2022]:[4] This model produces queries that have far fewer non-zero entries than the original Splade model, but documents that may have a larger number of non-zero entries.

References

P. Bajaj, D. Campos, N. Craswell, L. Deng, J. Gao, X. Liu, R. Majumder, A. McNamara, B. Mitra, T. Nguyen, M. Rosenberg, X. Song, A. Stoica, S. Tiwary, and T. Wang. Ms marco: A human generated machine reading comprehension dataset, 2018.

[1] Available at `https://huggingface.co/sentence-transformers/all-MiniLM-L6-v2`

[2] Available at `https://huggingface.co/sentence-transformers/msmarco-distilbert-base-tas-b`

[3] Pre-trained checkpoint from HuggingFace available at `https://huggingface.co/naver/splade-cocondenser-ensembledistil`

[4] Pre-trained checkpoints for document and query encoders were obtained from `https://huggingface.co/naver/efficient-splade-V-large-doc` and `https://huggingface.co/naver/efficient-splade-V-large-query`, respectively.

J. Devlin, M.-W. Chang, K. Lee, and K. Toutanova. BERT: Pre-training of deep bidirectional transformers for language understanding. In *Proceedings of the 2019 Conference of the North American Chapter of the Association for Computational Linguistics: Human Language Technologies, Volume 1 (Long and Short Papers)*, pages 4171–4186, June 2019.

T. Formal, C. Lassance, B. Piwowarski, and S. Clinchant. From distillation to hard negative sampling: Making sparse neural ir models more effective. In *Proceedings of the 45th International ACM SIGIR Conference on Research and Development in Information Retrieval*, page 2353–2359, 2022.

S. Hofstätter, S.-C. Lin, J.-H. Yang, J. Lin, and A. Hanbury. Efficiently teaching an effective dense retriever with balanced topic aware sampling. In *Proceedings of the 44th International ACM SIGIR Conference on Research and Development in Information Retrieval*, page 113–122, 2021.

O. Khattab and M. Zaharia. Colbert: Efficient and effective passage search via contextualized late interaction over bert. In *Proceedings of the 43rd International ACM SIGIR Conference on Research and Development in Information Retrieval*, pages 39–48, 2020.

C. Lassance and S. Clinchant. An efficiency study for splade models. In *Proceedings of the 45th International ACM SIGIR Conference on Research and Development in Information Retrieval*, page 2220–2226, 2022.

D. G. Lowe. Distinctive image features from scale-invariant keypoints. *International Journal of Computer Vision*, 60:91–110, 2004.

A. Oliva and A. Torralba. Modeling the shape of the scene: A holistic representation of the spatial envelope. *International Journal of Computer Vision*, 42:145–175, 2001.

J. Pennington, R. Socher, and C. Manning. GloVe: Global vectors for word representation. In *Proceedings of the 2014 Conference on Empirical Methods in Natural Language Processing*, pages 1532–1543, Oct. 2014.

N. Thakur, N. Reimers, A. Rücklé, A. Srivastava, and I. Gurevych. BEIR: A heterogeneous benchmark for zero-shot evaluation of information retrieval models. In *35th Conference on Neural Information Processing Systems Datasets and Benchmarks Track (Round 2)*, 2021.

Y. Wu, M. Schuster, Z. Chen, Q. V. Le, M. Norouzi, W. Macherey, M. Krikun, Y. Cao, Q. Gao, K. Macherey, J. Klingner, A. Shah, M. Johnson, X. Liu, Ł. Kaiser, S. Gouws, Y. Kato, T. Kudo, H. Kazawa, K. Stevens, G. Kurian, N. Patil, W. Wang, C. Young, J. Smith, J. Riesa, A. Rudnick, O. Vinyals, G. Corrado, M. Hughes, and J. Dean. Google's neural machine translation system: Bridging the gap between human and machine translation, 2016.

A. B. Yandex and V. Lempitsky. Efficient indexing of billion-scale datasets of deep descriptors. In *2016 IEEE Conference on Computer Vision and Pattern Recognition*, pages 2055–2063, 2016.

H. Zhang, X. Song, C. Xiong, C. Rosset, P. N. Bennett, N. Craswell, and S. Tiwary. Generic intent representation in web search. In *Proceedings of the 42nd International ACM SIGIR Conference on Research and Development in Information Retrieval*, pages 65–74, 2019.

Appendix B
Probability Review

Abstract We briefly review key concepts in probability in this appendix.

B.1 Probability

We identify a *probability space* denoted by $(\Omega, \mathcal{F}, \mathbb{P})$ with an *outcome space*, an *events* set, and a *probability measure*. The outcome space, Ω, is the set of all possible outcomes. For example, when flipping a two-sided coin, the outcome space is simply $\{0, 1\}$. When rolling a six-sided die, it is instead the set $[6] = \{1, 2, \ldots, 6\}$.

The events set $\mathcal{F}$ is a set of subsets of Ω that includes Ω as a member and is closed under complementation and countable unions. That is, if $E \in \mathcal{F}$, then we must have that $E^{\mathsf{C}}\mathcal{F}$. Furthermore, the union of countably many events E_i's in $\mathcal{F}$ is itself in $\mathcal{F}$: $\cup_i E_i \in \mathcal{F}$. A set $\mathcal{F}$ that satisfies these properties is called a σ-algebra.

Finally, a function $\mathbb{P} : \mathcal{F} \rightarrow \mathbb{R}$ is a probability measure if it satisfies the following conditions: $\mathbb{P}[\Omega] = 1$; $\mathbb{P}[E] \geq 0$ for any event $E \in \mathcal{F}$; $\mathbb{P}[E^{\mathsf{C}}] = 1 - \mathbb{P}[E]$; and, finally, for countably many disjoint events E_i's: $\mathbb{P}[\cup_i E_i] = \sum_i \mathbb{P}[E_i]$.

We should note that, $\mathbb{P}$ is also known as a "probability distribution" or simply a "distribution." The pair $(\Omega, \mathcal{F})$ is called a *measurable space*, and the elements of $\mathcal{F}$ are known as a *measurable sets*. The reason they are called "measurable" is because they can be "measured" with $\mathbb{P}$: The function $\mathbb{P}$ assigns values to them.

In many of the discussions throughout this monograph, we omit the outcome space and events set because that information is generally clear from context. However, a more formal treatment of our arguments requires a complete definition of the probability space.

S. Bruch, *Foundations of Vector Retrieval*, https://doi.org/10.1007/978-3-031-55182-6_12

B.2 Random Variables

A random variable on a measurable space $(\Omega, \mathcal{F})$ is a measurable function $X : \Omega \to \mathbb{R}$. It is measurable in the sense that the *preimage* of any Borel set $B \in \mathcal{B}$ is an event: $X^{-1}(B) = \{\omega \in \Omega \mid X(\omega) \in B\} \in \mathcal{F}$.

A random variable X generates a σ-algebra that comprises of the preimage of all Borel sets. It is denoted by $\sigma(X)$ and formally defined as $\sigma(X) = \{X^{-1}(B) \mid B \in \mathcal{B}\}$.

Random variables are typically categorized as discrete or continuous. X is *discrete* when it maps Ω to a discrete set. In that case, its *probability mass function* is defined as $\mathbb{P}[X = x]$ for some x in its range. A *continuous* random variable is often associated with a probability *density* function, f_X, such that:

$$\mathbb{P}[a \leq X \leq b] = \int_a^b f_X(x)dx.$$

Consider, for instance, the following probability density function over the real line for parameters $\mu \in \mathbb{R}$ and $\sigma > 0$:

$$f(x) = \frac{1}{\sqrt{2\pi\sigma^2}} e^{-\frac{(x-\mu)^2}{2\sigma^2}}.$$

A random variable with the density function above is said to follow a Gaussian distribution with mean μ and variance σ^2, denoted by $X \sim \mathcal{N}(\mu, \sigma^2)$. When $\mu = 0$ and $\sigma^2 = 1$, the resulting distribution is called the standard Normal distribution.

Gaussian random variables have attractive properties. For example, the sum of two independent Gaussian random variables is itself a Gaussian variable. Concretely, $X_1 \sim \mathcal{N}(\mu_1, \sigma_1^2)$ and $X_2 \sim \mathcal{N}(\mu_2, \sigma_2^2)$, then $X_1 + X_2 \sim \mathcal{N}(\mu_1 + \mu_2, \sigma_1^2 + \sigma_2^2)$. The sum of the squares of m independent Gaussian random variables, on the other hand, follows a χ^2-distribution with m degrees of freedom.

B.3 Conditional Probability

Conditional probabilities give us a way to model how the probability of an event changes in the presence of extra information, such as partial knowledge about a random outcome. Concretely, if $(\Omega, \mathcal{F}, \mathbb{P})$ is a probability space and $A, B \in \mathcal{F}$ such that $\mathbb{P}[B] > 0$, then the *conditional probability* of A given the event B is denoted by $\mathbb{P}[A \mid B]$ and defined as follows:

$$\mathbb{P}[A \mid B] = \frac{\mathbb{P}[A \cap B]}{\mathbb{P}[B]}.$$

We use a number of helpful results concerning conditional probabilities in proofs throughout the monograph. One particularly useful inequality is what is known as the *union bound* and is stated as follows:

$$\mathbb{P}[\cup_i A_i] \leq \sum_i \mathbb{P}[A_i].$$

Another fundamental property is the law of total probability. It states that, for mutually disjoint events A_i's such that $\Omega = \cup A_i$, the probability of any event B can be expanded as follows:

$$\mathbb{P}[B] = \sum_i \mathbb{P}[B \mid A_i]\,\mathbb{P}[A_i].$$

This is easy to verify: the summand is by definition equal to $\mathbb{P}[B \cap A_i]$ and, considering the events $(B \cap A_i)$'s are mutually disjoint, their sum is equal to $\mathbb{P}[B \cap (\cup A_i)] = \mathbb{P}[B]$.

B.4 Independence

Another tool that reflects the effect (or lack thereof) of additional knowledge on probabilities is the concept of *independence*. Two events A and B are said to be *independent* if $\mathbb{P}[A \cap B] = \mathbb{P}[A] \times \mathbb{P}[B]$. Equivalently, we say that A is independent of B if and only if $\mathbb{P}[A \mid B] = \mathbb{P}[A]$ when $\mathbb{P}[B] > 0$.

Independence between two random variables is defined similarly but requires a bit more care. If X and Y are two random variables and $\sigma(X)$ and $\sigma(Y)$ denote the σ-algebras generated by them, then X is independent of Y if all events $A \in \sigma(X)$ and $B \in \sigma(Y)$ are independent.

When a sequence of random variables are *mutually* independent and are drawn from the same distribution (i.e., have the same probability density function), we say the random variables are drawn *iid*: independent and identically-distributed. We stress that *mutual* independence is a stronger restriction than *pairwise* independence: m events $\{E_i\}_{i=1}^m$ are mutually independent if $\mathbb{P}[\cap_i E_i] = \prod_i \mathbb{P}[E_i]$.

We typically assume that data and query points are drawn *iid* from some (unknown) distribution. This is a standard and often necessary assumption that eases analysis.

B.5 Expectation and Variance

The *expected value* of a discrete random variable X is denoted by $\mathbb{E}[X]$ and defined as follows:

$$\mathbb{E}[X] = \sum_x x\,\mathbb{P}[X = x].$$

When X is continuous, its expected value is based on the following Lebesgue integral:

$$\mathbb{E}[X] = \int_\Omega X d\mathbb{P}.$$

So when a random variable has probability density function f_X, its expected value becomes:

$$\mathbb{E}[X] = \int x f_X(x) dx.$$

For a *nonnegative* random variable X, it is sometimes more convenient to unpack $\mathbb{E}\,X$ as follows instead:

$$\mathbb{E}[X] = \int_0^\infty \mathbb{P}[X > x] dx.$$

A fundamental property of expectation is that it is a linear operator. Formally, $\mathbb{E}[X + Y] = \mathbb{E}[X] + \mathbb{E}[Y]$ for two random variables X and Y. We use this property often in proofs.

We state another important property for independent random variables that is easy to prove. If X and Y are independent, then $\mathbb{E}[XY] = \mathbb{E}[X]\,\mathbb{E}[Y]$.

The *variance* of a random variable is defined as follows:

$$\mathrm{Var}[X] = \mathbb{E}\left[(X - \mathbb{E}[X])^2\right] = \mathbb{E}[X]^2 - \mathbb{E}[X^2].$$

Unlike expectation, variance is not linear unless the random variables involved are independent. It is also easy to see that $\mathrm{Var}[aX] = a^2\,\mathrm{Var}[X]$ for a constant a.

B.6 Central Limit Theorem

The result known as the Central Limit Theorem is one of the most useful tools in probability. Informally, it states that the average of *iid* random variables with finite mean and variance converges to a Gaussian distribution. There are several variants of this result that extend the claim to, for example, independent but not identically distributed variables. Below we repeat the formal result for the *iid* case.

Theorem B.1 *Let X_i's be a sequence of n iid random variables with finite mean μ and variance σ^2. Then, for any $x \in \mathbb{R}$:*

$$\lim_{n\to\infty} \mathbb{P}\Bigg[\underbrace{\frac{(1/n\sum_{i=1}^{n} X_i) - \mu}{\sigma^2/n}}_{Z} \le x\Bigg] = \int_{-\infty}^{x} \frac{1}{\sqrt{2\pi}} e^{-\frac{t^2}{2}}\, dt,$$

implying that $Z \sim \mathcal{N}(0, 1)$.

Appendix C
Concentration of Measure

Abstract By the strong law of large numbers, we know that the average of a sequence of m *iid* random variables with mean μ converges to μ with probability 1 as m tends to infinity. But how far is that average from μ when m is finite? Concentration inequalities helps us answer that question quantitatively. This appendix reviews important inequalities that are used in the proofs and arguments throughout this monograph.

C.1 Markov's Inequality

Lemma C.1 *For a nonnegative random variable X and a nonnegative constant $a \geq 0$:*

$$\mathbb{P}[X \geq a] \leq \frac{\mathbb{E}[X]}{a}.$$

Proof. Recall that the expectation of a nonnegative random variable X can be written as:

$$\mathbb{E}[X] = \int_0^\infty \mathbb{P}[X \geq x] dx.$$

Because $\mathbb{P}[X \geq x]$ is monotonically nonincreasing, we can expand the above as follows to complete the proof:

$$\mathbb{E}[X] \geq \int_0^a \mathbb{P}[X \geq x] dx \geq \int_0^a \mathbb{P}[X \geq a] dx = a\, \mathbb{P}[X \geq a].$$

□

S. Bruch, *Foundations of Vector Retrieval*, https://doi.org/10.1007/978-3-031-55182-6_13

C.2 Chebyshev's Inequality

Lemma C.2 *For a random variable X and a constant $a > 0$:*

$$\mathbb{P}\Big[|X - \mathbb{E}[X]| \geq a\Big] \leq \frac{\mathrm{Var}[X]}{a^2}.$$

Proof.

$$\mathbb{P}\Big[|X - \mathbb{E}[X]| \geq a\Big] = \mathbb{P}\Big[\big(X - \mathbb{E}[X]\big)^2 \geq a^2\Big] \leq \frac{\mathrm{Var}[X]}{a^2},$$

where the last step follows by the application of Markov's inequality. □

Lemma C.3 *Let $\{X_i\}_{i=1}^n$ be a sequence of iid random variables with mean $\mu < \infty$ and variance $\sigma^2 < \infty$. For $\delta \in (0, 1)$, with probability $1 - \delta$:*

$$\Big|\frac{1}{n}\sum_{i=1}^n X_i - \mu\Big| \leq \sqrt{\frac{\sigma^2}{\delta n}}.$$

Proof. By Lemma C.2, for any $a > 0$:

$$\mathbb{P}\Bigg[\Big|\frac{1}{n}\sum_{i=1}^n X_i - \mu\Big| \geq a\Bigg] \leq \frac{\sigma^2/n}{a^2}.$$

Setting the right-hand-side to δ, we obtain:

$$\frac{\sigma^2}{na^2} = \delta \implies a = \sqrt{\frac{\sigma^2}{\delta n}},$$

which completes the proof. □

C.3 Chernoff Bounds

Lemma C.4 *Let $\{X_i\}_{i=1}^n$ be independent Bernoulli variables with success probability p_i. Define $X = \sum_i X_i$ and $\mu = \mathbb{E}[X] = \sum_i p_i$. Then:*

$$\mathbb{P}\Big[X > (1 + \delta)\mu\Big] \leq e^{-h(\delta)\mu},$$

where,

$$h(t) = (1 + t)\log(1 + t) - t.$$

Proof. Using Markov's inequality of Lemma C.1 we can write the following for any $t > 0$:

$$\mathbb{P}\Big[X > (1+\delta)\mu\Big] = \mathbb{P}\Big[e^{tX} > e^{t(1+\delta)\mu}\Big] \leq \frac{\mathbb{E}\left[e^{tX}\right]}{e^{t(1+\delta)\mu}}.$$

Expanding the expectation, we obtain:

$$\begin{aligned}
\mathbb{E}\left[e^{tX}\right] &= \mathbb{E}\Big[e^{t\sum_i X_i}\Big] = \mathbb{E}\Big[\prod_i e^{tX_i}\Big] = \prod_i \mathbb{E}[e^{tX_i}] \\
&= \prod_i \Big(p_i e^t + (1-p_i)\Big) \\
&= \prod_i (1 + p_i(e^t - 1)) \\
&\leq \prod_i e^{p_i(e^t-1)} = e^{(e^t-1)\mu}. && \text{by } (1+t \leq e^t)
\end{aligned}$$

Putting all this together gives us:

$$\mathbb{P}\Big[X > (1+\delta)\mu\Big] \leq \frac{e^{(e^t-1)\mu}}{e^{t(1+\delta)\mu}}. \tag{C.1}$$

This bound holds for any value $t > 0$, and in particular a value of t that minimizes the right-hand-side. To find such a t, we may differentiate the right-hand-side, set it to 0, and solve for t to obtain:

$$\begin{aligned}
\frac{\mu e^t e^{(e^t-1)\mu}}{e^{t(1+\delta)\mu}} - \mu(1+\delta)\frac{e^{(e^t-1)\mu}}{e^{t(1+\delta)\mu}} &= 0 \\
\implies \mu e^t &= \mu(1+\delta) \\
\implies t &= \log(1+\delta).
\end{aligned}$$

Substituting t into Equation (C.1) gives the desired result. □

C.4 Hoeffding's Inequality

We need the following result, known as Hoeffding's Lemma, to present Hoeffding's inequality.

Lemma C.5 *Let X be a zero-mean random variable that takes values in $[a, b]$. For any $t > 0$:*

$$\mathbb{E}\left[e^{tX}\right] \leq \exp\Big(\frac{t^2(b-a)^2}{8}\Big).$$

Proof. By convexity of e^{tx} and given $x \in [a, b]$ we have that:

$$e^{tx} \leq \frac{b-x}{b-a}e^{ta} + \frac{x-a}{b-a}e^{tb}.$$

Taking the expectation of both sides, we arrive at:

$$\mathbb{E}\left[e^{tx}\right] \leq \frac{b}{b-a}e^{ta} - \frac{a}{b-a}e^{tb}.$$

To conclude the proof, we first write the right-hand-side as $\exp(h(t(b-a)))$ where:

$$h(x) = \frac{a}{b-a}x + \log\left(\frac{b}{b-a} - \frac{a}{b-a}e^{x}\right).$$

By expanding $h(x)$ using Taylor's theorem, it can be shown that $h(x) \leq x^2/8$. That completes the proof. □

We are ready to present Hoeffding's inequality.

Lemma C.6 *Let $\{X_i\}_{i=1}^n$ be a sequence of iid random variables with finite mean μ and suppose $X_i \in [a, b]$ almost surely. For all $\epsilon > 0$:*

$$\mathbb{P}\left[\left|\frac{1}{n}\sum_{i=1}^{n} X_i - \mu\right| > \epsilon\right] \leq 2\exp\left(-\frac{2n\epsilon^2}{(b-a)^2}\right).$$

Proof. Let $X = 1/n\sum_i X_i - \mu$. Observe by Markov's inequality that:

$$\mathbb{P}[X \geq \epsilon] = \mathbb{P}\left[e^{tX} \geq e^{t\epsilon}\right] \leq e^{-t\epsilon}\,\mathbb{E}[e^{tX}].$$

By independence of X_i's and the application of Lemma C.5:

$$\begin{aligned}
\mathbb{E}[e^{tX}] &= \mathbb{E}\left[\prod_i e^{\frac{t(X_i-\mu)}{n}}\right] \\
&= \prod_i \mathbb{E}\left[e^{\frac{t(X_i-\mu)}{n}}\right] \\
&\leq \prod_i \exp\left(\frac{t^2(b-a)^2}{8n^2}\right) \\
&= \exp\left(\frac{t^2(b-a)^2}{8n}\right).
\end{aligned}$$

We have shown that:

$$\mathbb{P}[X \geq \epsilon] \leq \exp\left(-t\epsilon + \frac{t^2(b-a)^2}{8n}\right).$$

That statement holds for all values of t and in particular one that minimizes the right-hand-side. Solving for that value of t gives us $t = 4n\epsilon/(b-a^2)$, which implies:

$$\mathbb{P}[X \geq \epsilon] \leq e^{-\frac{2n\epsilon^2}{(b-a)^2}}.$$

By a symmetric argument we can bound $\mathbb{P}[X \leq -\epsilon]$. The claim follows by the union bound over the two cases. □

C.5 Bennet's Inequality

Lemma C.7 *Let $\{X_i\}_{i=1}^n$ be a sequence of independent random variables with zero mean and finite variance σ_i^2. Assume that $|X_i| \leq a$ almost surely for all i. Then:*

$$\mathbb{P}\left[\sum_i X_i \geq t\right] \leq \exp\left(-\frac{\sigma^2}{a^2} h\left(\frac{at}{\sigma^2}\right)\right),$$

where $h(x) = (1+x)\log(1+x) - x$ and $\sigma^2 = \sum_i \sigma_i^2$.

Proof. As usual, we take advantage of Markov's inequality to write:

$$\begin{aligned}
\mathbb{P}\left[\sum_i X_i \geq t\right] &\leq e^{-\lambda t}\,\mathbb{E}\left[e^{\lambda \sum_i X_i}\right] \\
&= e^{-\lambda t}\,\mathbb{E}\left[\prod_i e^{\lambda X_i}\right] \\
&= e^{-\lambda t}\prod_i \mathbb{E}\left[e^{\lambda X_i}\right]
\end{aligned}$$

Using the Taylor expansion of e^x, we obtain:

$$\begin{aligned}
\mathbb{E}\left[e^{\lambda X_i}\right] &= \mathbb{E}\left[\sum_{k=0}^{\infty} \frac{\lambda^k X_i^k}{k!}\right] \\
&= 1 + \sum_{k=2}^{\infty} \frac{\lambda^k\,\mathbb{E}[X_i^2 X_i^{k-2}]}{k!} \\
&\leq 1 + \sum_{k=2}^{\infty} \frac{\lambda^k \sigma_i^2 a^{k-2}}{k!} \\
&= 1 + \frac{\sigma_i^2}{a^2} \sum_{k=2}^{\infty} \frac{\lambda^k a^k}{k!} \\
&= 1 + \frac{\sigma_i^2}{a^2}\left(e^{\lambda a} - 1 - \lambda a\right) \\
&\leq \exp\left(\frac{\sigma_i^2}{a^2}\left(e^{\lambda a} - 1 - \lambda a\right)\right).
\end{aligned}$$

Putting it all together:

$$\mathbb{P}\left[\sum_i X_i \geq t\right] \leq e^{-\lambda t} \prod_i \exp\left(\frac{\sigma_i^2}{a^2}\left(e^{\lambda a} - 1 - \lambda a\right)\right)$$
$$= e^{-\lambda t} \exp\left(\frac{\sigma^2}{a^2}\left(e^{\lambda a} - 1 - \lambda a\right)\right).$$

This inequality holds for all values of λ, and in particular one that minimizes the right-hand-side. Setting the derivative of the right-hand-side to 0 and solving for λ leads to the desired result. □

Appendix D
Linear Algebra Review

Abstract This appendix reviews basic concepts from Linear Algebra that are useful in digesting the material in this monograph.

D.1 Inner Product

Denote by $\mathbb{H}$ a vector space. An inner product $\langle \cdot, \cdot \rangle : \mathbb{H} \times \mathbb{H} \rightarrow \mathbb{R}$ is a function with the following properties:

- $\forall\, u \in \mathbb{H},\ \langle u, u \rangle \geq 0;$
- $\forall\, u \in \mathbb{H},\ \langle u, u \rangle = 0 \Leftrightarrow u = 0;$
- $\forall\, u, v \in \mathbb{H},\ \langle u, v \rangle = \langle v, u \rangle$; and,
- $\forall\, u, v, w \in \mathbb{H},\ \textit{and}\ \alpha, \beta \in \mathbb{R},\ \langle \alpha u + \beta v, w \rangle = \alpha \langle u, w \rangle + \beta \langle v, w \rangle.$

We call $\mathbb{H}$ together with the inner product $\langle \cdot, \cdot \rangle$ an *inner product space*. As an example, when $\mathbb{H} = \mathbb{R}^d$, given two vectors $u = \sum_{i=1}^d u_i e_i$ and $v = \sum_{i=1}^d v_i e_i$, where e_i's are the standard basis vectors, the following is an inner product:

$$\langle u, v \rangle = \sum_{i=1}^{d} u_i v_i.$$

We say two vectors u and v in an inner product space are *orthogonal* if their inner product is 0: $\langle u, v \rangle = 0$.

D.2 Norms

A function $\Phi : \mathbb{H} \rightarrow \mathbb{R}_+$ is a norm on $\mathbb{H}$ if it has the following properties:

- Definiteness: For all $u \in \mathbb{H}$, $\Phi(u) = 0 \Leftrightarrow u = 0;$

S. Bruch, *Foundations of Vector Retrieval*, https://doi.org/10.1007/978-3-031-55182-6_14

- Homogeneity: For all $u \in \mathbb{H}$ and $\alpha \in \mathbb{R}$, $\Phi(\alpha u) = |\alpha|\Phi(u)$; and,
- Triangle inequality: $\forall\, u, v \in \mathbb{H},\ \Phi(u+v) \leq \Phi(u) + \Phi(v)$.

Examples include the absolute value on $\mathbb{R}$, and the L_p norm (for $p \geq 1$) on $\mathbb{R}^d$ denoted by $\|\cdot\|_p$ and defined as:

$$\|u\|_p = \Big(\sum_{i=1}^{d} |u_i|^p\Big)^{\frac{1}{p}}.$$

Instances of L_p include the commonly used L_1, L_2 (Euclidean), and L_∞ norms, where $\|u\|_\infty = \max_i |u_i|$.

Note that, when $\mathbb{H}$ is an inner product space, then the function $\|u\| = \sqrt{\langle u, u\rangle}$ is a norm.

D.3 Distance

A norm on a vector space induces a notion of distance between two vectors. Concretely, if $\mathbb{H}$ is a normed space equipped with $\|\cdot\|$, then we define the distance between two vectors $u, v \in \mathbb{H}$ as follows:

$$\delta(u, v) = \|u - v\|.$$

D.4 Orthogonal Projection

Lemma D.1 *Let $\mathbb{H}$ be an inner product space and suppose $u \in \mathbb{H}$ and $u \neq 0$. Any vector $v \in \mathbb{H}$ can be uniquely decomposed along u as:*

$$v = v_\perp + v_\parallel,$$

such that $\langle v_\perp, v_\parallel \rangle = 0$. Additionally:

$$v_\parallel = \frac{\langle u, v\rangle}{\langle u, u\rangle} u,$$

and $v_\perp = v - v_\parallel$.

Proof. Let $v_\parallel = \alpha u$ and $v_\perp = v - v_\parallel$. Because $v_\parallel$ and $v_\perp$ are orthogonal, we deduce that:

$$\langle v_\parallel, v_\perp \rangle = 0 \implies \langle \alpha u, v_\perp \rangle = 0 \implies \langle u, v_\perp \rangle = 0.$$

That implies:

$$\langle v, u \rangle = \alpha \langle u, u \rangle \implies \alpha = \frac{\langle u, v \rangle}{\langle u, u \rangle},$$

so that:

$$v_{\|} = \frac{\langle u, v \rangle}{\langle u, u \rangle} u.$$

We prove the uniqueness of the decomposition by contradiction. Suppose there exists another decomposition of v to $v'_{\|} + v'_{\perp}$. Then:

$$\begin{aligned} v_{\|} + v_{\perp} = v'_{\|} + v'_{\perp} &\implies \langle u, v_{\|} + v_{\perp} \rangle = \langle u, v'_{\|} + v'_{\perp} \rangle \\ &\implies \langle u, v_{\|} \rangle = \langle u, v'_{\|} \rangle \\ &\implies \langle u, \alpha u \rangle = \langle u, \beta u \rangle \\ &\implies \alpha = \beta. \end{aligned}$$

We must therefore also have that $v_{\perp} = v'_{\perp}$. □

Printed in the United States
by Baker & Taylor Publisher Services